FORAGE CONSERVATION AND FEEDING

FRANK RAYMOND
PETER REDMAN
RICHARD WALTHAM

Illustrations by
CHARLES RAYMOND
CHRISTOPHER RAYMOND

FARMING PRESS LTD
Wharfedale Road, Ipswich, Suffolk IP1 4LG

First published 1972
Fourth Edition (completely revised and reset) 1986

British Library Cataloguing in Publication Data

Raymond, Frank
  Forage conservation and feeding.
  4th ed., completely rev. and reset
  1. Forage plants   2. Silage
  I. Title   II. Redman, Peter   III. Waltham,
  Richard
  633.2′086   SB195

  ISBN 0-85236-139-4

Cover design by Hannah Berridge

Phototypeset by Galleon Photosetting, Ipswich
Printed and bound in Great Britain by Hazell Watson & Viney Limited,
Member of the BPCC Group, Aylesbury, Bucks

Contents

PAGE

1 INTRODUCTION   1

2 THE PRINCIPLES OF FORAGE CONSERVATION   8

Haymaking—barn hay-drying—chemical additives.
High-temperature drying.
Ensilage—wilting—additives.

3 THE FEEDING VALUE OF CONSERVED FORAGES   27

Digestibility—forages at cutting—conserved forage—D-values and metabolis-
able energy.
Feed intake—differences between forage species—dried grass—silage—increas-
ing silage intake—other factors limiting intake.
Conserved forages fed in mixed rations—balanced rations—conserved forages
and cereals—protein value of conserved forages—fibre in ruminant rations
—mineral content of conserved forages.

4 CROPS FOR CONSERVATION   46

Digestibility levels—crops for hay—crops for silage—manuring grass and
forage crops—forage maize—kale and crop by-products for silage—crops for
grass-drying.

5 MOWING AND SWATH TREATMENT   63

Basic requirements—treatment after mowing.
Types of mowing and conditioning equipment—cutter-bar mowers—rotary
drum and multiple disc mowers—combined equipment—flail mowers—
mower-conditioners.
Operating equipment—swath-handling machinery—crop loss and drying rate—
non-mechanical conditioning.

6 HAYMAKING   84

Bales and balers—standard balers—handling standard bales—loading into
store—overall system performance—large bales.
Barn hay-drying—systems—choice of driers—drying large rectangular bales—
handling and drying loose crops.

7   SILAGE MAKING   105

Harvesting the crop—stage of growth at cutting—dry-matter content—chop length—types of harvester—forage harvesting systems—self-loading forage wagons—harvesting maize for silage—avoiding soil contamination.
Filling the silo: Dorset wedge method—walled silos—outdoor silos—clamp silos—ensiling in a polythene sleeve—big-bale silage.
Other aspects—ensiling forage maize—the use of silage additives—silage with reduced fermentation—effluent loss—tower silos—care during storage of silage.

8   STRAW AS ANIMAL FEED   141

The digestibility of straw—collection and storage—treating straw.

9   METHODS OF FEEDING   150

Hay and straw.
Silage—self-feeding—mechanised feeding—silo unloaders—block cutters—self-loading feeders—forage boxes—mixer feeder wagons—feed troughs—feeding big-bale silage—mechanised feeding from tower silos.

10   FEEDING CONSERVED FORAGES   163

Dairy cow feeding—conserved forages for beef cattle—sheep.

11   FORAGE CONSERVATION IN FARMING SYSTEMS   178

INDEX   185

Preface to the Fourth Edition

IT IS now fourteen years since this book was first published. Despite considerable amendments and additions in subsequent editions it was clear to the editors that much of its content was beginning to be outdated, and that a major revision was needed. This we have aimed to do in this Fourth Edition.

Co-author with us in the first three editions was our friend and colleague, Gordon Shepperson. Sadly Gordon died in 1979, and practical agricultural science is the poorer. We know that Gordon Shepperson would have actively supported the present revision; we are certain too that, with us, he would have welcomed his colleague Peter Redman as joint author in this revised edition.

April 1986.

FRANK RAYMOND
RICHARD WALTHAM

Chapter 1

INTRODUCTION

FORAGE CONSERVATION AND FEEDING was first published in 1972, just before the United Kingdom became a member of the European Economic Community. At that time it was still necessary to *argue* the case for better forage conservation, for official policy appeared to be to subsidise cereals and concentrate feeds at prices low enough for them to be fed profitably for the whole of the production ration. In fact, in the early 1960s cereals were cheap enough to make up part of the maintenance ration fed to livestock, as in straw-balancer milk systems and barley beef production. A particular consequence was that the low price of cereals, relative to returns from milk and beef, tended to discourage better grassland management. Keen grassland farmers did of course seek to make good hay and silage—and could show lower winter feed costs as a result. But in many cases their management problems would have been easier—and they would have made more money—if they had increased their summer stocking rates, conserved less grass, and purchased more subsidised cereals and concentrates.

A quite new situation opened up during the period 1972 to 1975, as a result of EEC entry, combined with the effects of the first 'oil crisis' and of a sudden world shortage of cereals and protein feeds following a series of poor harvests in the main producing countries. Together, these events encouraged farmers, in Europe as well as in the United Kingdom, to speed up the rate at which they adopted the new technologies which were just beginning to emerge from the expanded R & D programmes of the post-War years. These were quickly to transform food shortage into food surplus, not through more animals being fed or more land being cultivated, but by raising output per animal and yield per hectare. The marked increase in milk yield per cow that occurred from the early 1970s (Figure 1.1) was due to the combined effects of better breeding through AI, improved health, cubicle housing instead of stalls, and better feeding; as a result milk output per cow began to increase at 4 per cent a year, after twenty years during which it had increased at only 2 per cent a year. Similarly the combination of higher-yielding varieties,

1

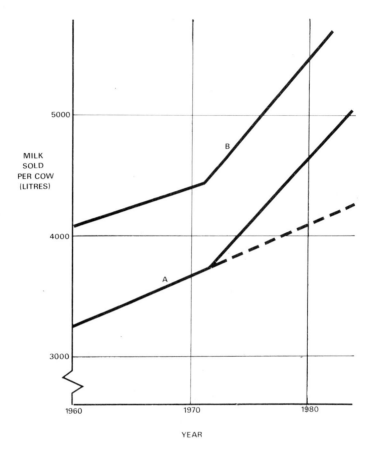

Figure 1.1   Line A: Average annual milk yield of dairy cows in the United Kingdom. Line B: Average of 2300 higher-yielding herds recorded by BOCM-Silcock

higher fertiliser use, fungicides which allowed winter varieties to be grown, and herbicides to control wild oats and blackgrass, raised average wheat yields from 4 to 6 tonnes per hectare over the same period.

Higher concentrate feeding contributed much to the higher milk yields; but so did forage conservation, as dairy farmers adopted better methods. The most marked change was in the shift from haymaking to silage as the main method of conservation. As Figure 1.2 shows, in 1970 only about 10 per cent of the forage conserved in the UK was in the form of silage; by the mid-1980s this had risen to over 70 per cent, with the most marked swing to silage being on dairy farms (Table 1.1).

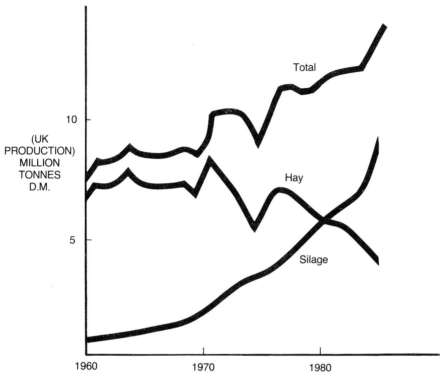

Figure 1.2   Quantities of hay and silage conserved in the United Kingdom, 1960–84. The data are not exact as they are based on on-farm estimates of hay and silage made

The risk of surplus milk production in the EEC had been foreseen by Mansholt back in the 1960s, but was largely forgotten during the 'crises' of the early 1970s. But rising milk production per cow soon began to make surplus a reality, so that by the 1980s Europe had increasing amounts of unsaleable butter and skim-milk powder in Intervention

Table 1.1   Percentage of grass fields cut one or more times for hay and silage in dairying and beef and sheep farming regions in England and Wales

	Dairying		*Beef and sheep*	
	1974	1984	1974	1984
Hay	28	17	27	18
Silage	18	33	6	22

(Data: Archer, ADAS)

stores. The classical solution to overproduction had been to reduce prices, and between 1976 and 1984 the 'real' price paid for milk (that is, allowing for inflation) fell by over 20 per cent. But this proved ineffective, because Intervention buying offered a virtually unlimited market for milk, and dairy farmers were able to compensate for the lower prices by producing, and selling, more milk. They did this by intensifying their production, in particular by feeding more concentrates to each cow.

By 1984 it had become evident that price cuts alone, at least at any politically acceptable level, would not be enough to curb surplus milk production, and milk quotas were introduced. Quotas meant that each dairy farmer in the UK could now sell only 93 per cent of the milk he had sold in the previous year; no longer could he compensate for price cuts by selling more milk.

Introduction of quotas clearly presented problems, particularly for the Milk Boards. But they did not prove the disaster for very many dairy farmers that had perhaps been predicted. Instead they led to many more dairy farmers adopting feeding systems based on greater use of grazed and conserved forages, which had until then only been operated by grassland enthusiasts. There are clearly risks in comparing just two years. But several national studies comparing dairy cow feeding in 1984, the first year of quotas, with the previous year indicate the response that was made (e.g. Table 1.2). Despite the reduction in the amount of milk sold, as well as a small fall in the price received per litre, margins per cow and per hectare if anything rose, *as farmers sought to produce their quota as cheaply as possible.* The methods they adopted were not new; there had just not been sufficient incentive to use them before because there had been no limit on the amount of milk that could be sold.

Table 1.2   Comparison of margins in milk production in 1983–4 with 1984–5
(the first year of milk quotas)

	1983–4	1984–5
Herd size	108	105
Milk yield per cow (*litres*)	5,452	5,317
Milk price (*pence/litre*)	14.8	14.5
Milk output per cow (£)	806	769
Concentrates fed per litre (*kg*)	0.33	0.25
Margin over purchased feed (£)	511	564
Stocking rate (*cows/hectare*)	2.16	2.15
Margin (£/hectare)	1,015	1,115

(Data: MMB/ADAS Milkminder)

There is little doubt though that, in order for forages to replace some of the concentrates that had previously been fed, both a higher standard of management and a greater attention to detail were needed, in particular in making the increased amounts of hay and silage that were now required. As we have noted, the techniques for doing this were already largely to hand; it needed the discipline of milk quotas to encourage more farmers to adopt them.

There has probably been less pressure on beef producers to make better use of forages, at least partly because the cost of barley has remained fairly in line with beef prices, and there has so far been no restriction on beef production. Increasingly, though, the amount of beef produced is being limited by the supply of suitable calves, both because dairy cow numbers have fallen as a result of quotas and because the economics of beef breeding herds have remained marginal. High calf prices will continue and could encourage the feeding of beef cattle to higher weights, in line with some current trends in the market for beef. Clearly too there is a demand for leaner beef, in response both to consumer preference and to concern about healthy eating. These trends point towards the use of more forages in beef feeding, and less use of cereals, which tend to 'finish' cattle at lower weights and to produce fat in the carcass. However, prediction is difficult. The proposed EEC ban on growth promoters, which are widely used in beef production; the feeding of bulls, which grow faster and convert their feed better than steers; and the feeding of more heifers for beef, will all influence the optimum rations that are fed for beef production.

The position seems clearer with sheep, where the good returns resulting from the EEC sheepmeat regime have encouraged the wide adoption of indoor winter housing, with its benefits in reduced mortality of ewes and lambs, higher lamb birth weights, and improved conditions for shepherding. Housing has, however, greatly increased the winter requirements for conserved forage, and there have already been important developments, particularly in making and feeding silage to housed sheep. Taking sheep off pastures in winter has also undoubtedly improved summer grass production, both for grazing, and in making more grass available for cutting.

These pressures for better forage conservation have already led to major changes from the situation described in 1972. The shift from hay to silage has made it more practicable to cut crops at a less mature stage, and so of potentially higher feeding value, and much greater care is now being taken to prevent losses—as is evident from the large numbers of well-sealed silos that can be seen round the countryside every autumn (Plate 1.1). But there is still much scope for further improvement; during the wet summer of 1985 large areas of hay were left to deteriorate

Plate 1.1 *Typical outdoor clamp silo covered with closely packed tyres to prevent surface wastage and damage during storage (Photograph* Dairy Farmer*).*

for weeks in the field because the hay was too wet to bale, and much badly fermented wet silage was made.

But attitudes have also changed. In 1972 we commented that developments in forage conservation were being held back because the concept that hay and silage were 'maintenance' feeds—with concentrates fed for 'production'—meant that only simple and cheap methods of conservation were acceptable; that these tended to produce low-value feeds; and that such feeds could clearly only justify cheap methods. We argued that research on forage conservation should at least initially be concerned with biological efficiency—how to produce high-value feeds with low losses. More efficient methods would not necessarily turn out to be either complex or expensive; but they were most unlikely to be developed if the first consideration was cheapness.

Fortunately much research on better conservation methods was by then already underway. The application of that research has been one of the factors that has enabled UK farmers to respond so well to the increasing economic pressures on livestock production. Yet, as was predicted, the better conservation methods that are being used are not greatly more complex, or more expensive, than the less effective methods they have largely replaced. To be successful they do require the application of more knowledge and attention to detail than did earlier methods.

Thus much of what we wrote about in 1972 is now accepted, and applied, on livestock farms. But there is still much scope for further improvement, in particular to make forage conservation more independent of the weather and so more predictable. To achieve this, we believe, requires first a good understanding of the principles determining forage nutritive value, how losses in conservation occur, and how these losses may affect nutritive value; and secondly, how this information can be used to develop more efficient systems that can be applied under practical farm conditions. The progress that has already been made merely serves as a pointer to the great potential in conserved forages that still remains untapped. If this message was timely in 1972 it is, though perhaps for rather different reasons, equally so in 1986.

Chapter 2

THE PRINCIPLES OF FORAGE CONSERVATION

FRESH GREEN CROPS continue to 'live' for some time after they are cut. In fact in bright sunshine they may photosynthesise and even increase in dry weight for a short while after cutting. But soon the crop dies; the plant cells lose moisture, the sugars in the plant sap begin to oxidise, and the plant proteins to break down. At the same time bacteria and moulds, which are always present on the surface of the growing crop but which can generally do little damage to the living plant, begin to attack and decompose the dying tissues.

The aim of an efficient conservation process is to check these destructive processes rapidly and completely so as to preserve as much as possible of the yield and feeding value of the crop. Two main processes are used; either the crop is dried, by haymaking or high-temperature dehydration, to a stage at which both chemical and microbial action cease; or it is preserved at a high moisture content by the action of acids or other chemicals, in the process of ensilage. Some background understanding of these different processes is of great help both in choosing the most suitable method for a particular farm, and in carrying it out successfully under practical farming conditions.

HAYMAKING

Haymaking aims to produce a stable product of medium nutritive value with the minimum of wastage and loss and at a reasonable capital and labour cost. In the process between 70 and 95 per cent of the moisture present in the crop when it is cut is removed by the action of sun and wind as the crop lies in the field before it is removed for storage. The moisture content (MC) at which hay can be safely stored depends on the type of crop. Thus hay which is cut at a fairly mature stage of growth contains relatively little sugar which can oxidise and cause heating in storage, and is generally safe to store when its moisture content has been reduced to between 15 and 18 per cent. In contrast hay

8

made from young, leafy crops may contain a higher sugar content, and unless it is dried below 12 per cent MC it is likely to heat up and mould in store. Weather conditions rarely allow hay to be dried to this level in the field, and it is often necessary to arrange for some remaining moisture to be removed from the crop after it is picked up, preferably by some system of forced-air ventilation under cover of the barn. Even then there is a risk of hay made from very young crops picking up some atmospheric moisture during storage, with subsequent moulding.

Young grass cut for quality hay is likely to have a moisture content of 80 per cent or more when it is cut; this means that to produce one tonne of dry hay (containing say 15 per cent MC at the time it is fed) 3.25 tonnes of water have to be removed. If the crop is allowed to become more mature before it is cut its moisture content may well have decreased to 75 per cent, at which level only 2.4 tonnes of water have to be removed to produce a tonne of 'dry' hay. But while the more mature crop can be made into hay more quickly, and probably with lower losses, the hay will tend to be of lower feeding value. The best balance to aim for between product quality and ease and certainty of making is the major decision that has to be made by the farmer planning to make hay.

Generally the aim will be to reduce the moisture content of a cut crop to below 25 per cent before it is baled. At this level only about a further 150 kg of moisture, out of the original 3 tonnes or so in the crop at the time of cutting, have to be removed to produce a tonne of dry hay, safe for storage. With many crops this can be done by making small stacks of bales in the field to allow further drying as well as dissipation of any heat produced by continued respiration; but as already noted, with immature crops, or with hay baled at a higher moisture content, either further artificial drying, or treatment with chemicals, is necessary to prevent deterioration and loss of nutritive value during storage.

For a good hay crop therefore between 15 and 20 tonnes of water have to be evaporated for each hectare of crop harvested. Three-quarters of this water can be lost on the day that the crop is mown, given suitable treatment and reasonable weather, because water is readily lost from the outer surfaces of plants, down to about 65 per cent MC. However, this will only happen in practice if the swath of cut crop is opened up so that drying air can penetrate it. In contrast a tightly packed swath will dry very slowly; in fact in poor drying weather its moisture content may increase, even in the absence of rain, because additional water is being produced by oxidation of 'sugars' in the plant material.

Thus immediate action must be taken to 'open up' the crop, by tedding for example, to allow the water to evaporate. Moisture content can also differ markedly between the top of the swath, exposed to drying air, and the bottom which may be lying on damp soil; thus tedding also encourages more uniform drying within the swath.

Once the crop has reached 60 per cent MC, with much of the moisture from the leaves and outer surfaces of the stems in the crop having been lost, the next stage of drying, down to about 30 per cent MC, is much slower because water moves increasingly slowly from within the stems and larger plant fractions. The rate and evenness of drying during this stage are greatly assisted by some form of mechanical conditioning, applied at the time of cutting, which aims to damage the surface of the plants and so increase rate of water loss.

In the final stages of drying, from 30 per cent MC, the rate of drying is greatly influenced by the prevailing weather conditions, and particularly by the humidity of the air. In all but the best conditions this stage of drying can be very protracted, and it is then that conservation methods which allow hay to be removed from the field into store before it is fully dry are especially beneficial.

In practice drying to less than 20 per cent MC in the *swath*—that is with no treatment of the crop after it is mown—is seldom possible, and even when it is achieved it can be at the cost of losing much of the leaf, and therefore much of the protein content, of the hay. In any event rain falling on nearly dry hay, whether or not it has been tedded or conditioned, causes severe loss of nutrients by leaching, and reduces nutritive value, in particular its energy content. If nearly dry hay is subjected to prolonged or repeated wetting its MC may well increase back to as high as 70 per cent, regardless of the type of swath or the treatment that has been applied. This rewetting can lead to moulding and rotting of the crop and, even if it is ultimately dried enough to bale safely, the hay made is likely to be of very poor quality.

Physical and leaching losses in the field can exceed 30 per cent of the dry matter under bad conditions. But unless the hay is fully dried there can then be further losses, both of dry matter and of nutritive value, resulting from continued respiration and microbial activity which can occur in field heaps and in store. Although rate of respiration—and so dry-matter loss—does decrease as the temperature in a heap of bales rises above 32°C, it continues until the plant cells start to die at about 45°C. But at higher temperatures the activity of bacteria and moulds can then cause a further loss of as much as 10 per cent of the dry matter in the original crop. Unless it is controlled temperature continues to rise still further, and at about 70°C chemical oxidation starts, with a rapid rise in temperature to over 200°C, and the risk of spontaneous combustion of the stack. Fortunately this degree of heating does not often occur in practice; but the problems of mouldy hay, which is unpalatable and can create health hazards, and of brown overheated hay, which, though palatable, is likely to be of low feeding value, are very common. It is estimated that in some years over 80 per cent of the hay made in the wetter parts of the country is either moulded or overheated.

The value of mechanical treatment of the cut crop in the field, so as to speed up the overall rate of water loss and to improve the evenness of drying between leaf and stem, has been emphasised as a way of reducing losses of dry matter and nutrients. But even with such treatments drying to a sufficiently low moisture content in the field is difficult because, as the hay approaches equilibrium with atmospheric moisture, the rate of water loss decreases; the stems remain succulent while the leaves, which dry more quickly, become brittle and can be lost. Consequently the decision to bale and remove the crop from the field becomes a compromise between accepting a moisture content which may be too high for safe keeping in store, and leaving the crop to dry completely in the field, with the risks of leaching and physical loss.

The Role of Barn Hay-Drying

The position may be partly remedied by grouping bales and leaving them to 'cure' in small heaps or stacks in the field. But a more certain way to prevent losses from heating and moulding in store is to stack the bales under cover and to finish drying by some form of fan ventilation, generally referred to as barn-drying (page 97). Table 2.1 indicates the moisture levels that must be reached for different systems of hay-making, including barn-drying, and the time the crop may have to

Table 2.1 Moisture contents at which hay can be removed from the field

Treatment	Moisture content limits (%)	Swath exposure time (hours)
Barn hay-drying, using some heated air	45–60	8–72
Baled, chopped or loose hay dried in a barn or tunnel without supplementary heat	35–40	24–96
Storage conditioning of baled or chopped hay	30–35	48–120
Hay treated with a chemical additive, e.g. propionic acid	25–35	48–120
Baling followed by field conditioning and barn storage	20–30	48 hours, up to 2 weeks or more

remain in the field for these levels to be reached under average conditions. But it must be emphasised that where a method is adopted which allows hay to be collected from the field at above 30 per cent MC, it is more than ever important to ensure that the different parts of the swath or windrow dry as evenly as possible, because variations in moisture content between different bales in a stack, or within the individual bales, can cause real problems in subsequent drying operations.

A number of different systems of barn-drying have been developed, but they have been slow to gain farmer acceptance, and although the advantages to be gained are not in doubt it is probable that less than 4 per cent of all hay in the United Kingdom is made in this way. Advantages that have been shown include higher yields of dry matter—up to 15 per cent greater than with conventional methods of hay-making—and the better feeding value of the hay produced. As Table 2.2 shows, this improved feeding value results both from the earlier cutting which is made practicable by barn-drying, and because more of the digestible nutrients and protein in the crop cut are retained, compared with hay made in the field.

Table 2.2   **The digestibility (D-value) and crude protein content of hay made by different methods from a crop cut at increasing stages of maturity**

Date of cutting	Cut grass		Barn-dried hay		Field-made hay	
(1958)	D-value	%CP	D-value	%CP	D-value	%CP
6 June	65.5	11.1	62.5	10.8	58.2*	9.5
17 June	60.5	9.5	52.0*	8.5	45.7*	8.4
25 June	55.2	7.5	52.2*	8.6	41.5*	7.8

*(Crop affected by rain in the field)
(Data: Shepperson, NIAE)

The main reasons for the slow acceptance of barn-drying during its early development during the 1950s were the low output and the need for double man-handling of bales in the 'batch' systems then in use; hence the system was most suited to small farms, making perhaps 30 tonnes of hay a year. Handling costs were also high because of lack of mechanisation, both with bales and in the loose-drying systems in which long unbaled hay had to be spread uniformly by hand. Poor design of driers and inadequate airflow also led to uneven and inefficient drying.

The low throughput of most driers also meant that cutting often became delayed well beyond the optimum crop growth stage; thus despite the added expense of handling and drying the feeding potential of much of the hay made was disappointingly low.

The Use of Chemical Additives in Haymaking

These practical difficulties have largely been overcome in modern systems of barn-drying, as described in Chapter 6. Yet there is still only limited adoption of the system, and little indication that it will play a major role in the future, so the problem remains that in most years haymaking by field-drying alone is a risky operation. Thus interest has been directed to the possible role of chemicals added to damp hay to prevent subsequent decomposition and moulding.

Under experimental conditions a number of additives, in which propionic acid has generally been the main ingredient, have been shown to prevent the development of moulds in hay containing up to 30 per cent of residual moisture. They act by preventing the continuation of plant respiration, which uses up some of the sugars remaining in the hay and at the same time produces heat and water which encourage moulding; more seriously, as the temperature of the hay rises above 40°C potentially harmful fungi may start to develop. These can produce toxic spores which, if inhaled, can cause Farmer's Lung in humans and mycotic abortion in cattle; propionic acid also has specific anti-fungal activity.

The earlier experimental work with hay additives was followed up in field trials which indicated that to prevent moulding under farm conditions about 6 kg of propionic acid needs to be retained in each tonne of hay at 30 per cent MC. However, propionic acid is volatile, and part of the acid used may be lost while it is being applied during baling. Thus to ensure safe storage the application rate may need to be doubled, up to 12 kg per tonne of moist hay. A salt of propionic acid, ammonium propionate, has the advantage of being less volatile; thus application losses are lower and it is consequently more pleasant to use. However, ammonium propionate contains only about 65 per cent of the activity of the acid in inhibiting mould growth so that proportionately more must be applied—about 6 kg per tonne of moist hay at 25 per cent MC, increasing to 12 kg at 30 per cent MC. These rates have proved effective with both conventional bales and large rectangular bales. However, large round bales require higher levels of additive; thus about 18 kg of ammonium propionate is needed with hay of 25 per cent MC. With round bales of higher moisture content it has been found difficult to prevent moulding regardless of the amount of additive used.

A major problem in the use of hay preservatives is in applying the liquid containing the chemical uniformly to the hay as it is baled. Several different application systems have been investigated, and these have shown the advantage of applying the chemical solution in large drops (700–1,200 mμ diameter) at, or very close to, the baler pick-up. These larger drops reduce the loss of volatile acid and are more likely to penetrate the crop mass. Distribution is achieved by the drops shattering on impact and by the mixing effect as the crop passes through the baler.

For effective and economic treatment the rate of chemical applied must be precisely related to crop moisture content and to baler through-put, both of which can vary considerably as a crop is harvested in the field. Thus application rate will need to be altered at intervals as crop condition and baling rate change during the day.

Many farmers now use some type of chemical preservative in an attempt to reduce the risk of deterioration of hay which is not completely dry. A number of commercial formulations are marketed; while these differ in composition, an additive is most likely to be effective in raising the safe baling moisture content of hay if the recommended application rate contains either propionic acid or ammonium propionate at the rates noted above. However, it is difficult to quantify the benefits of the use of preservatives, because they have been introduced at much the same time as the widespread adoption of important new developments in field conditioning, which have greatly speeded up the rate and extent of moisture loss from hay. In practice the combination of faster field drying with the tactical use of an additive when drying conditions are unfavourable has undoubtedly contributed to the improvement in hay quality that has occurred; the average protein content in hay increased by something over 0.5 per cent during the 1970s.

Current research is also studying the possibility of storing moist hay by treatment with caustic soda, aqueous ammonia or anhydrous ammonia (page 133). These chemicals appear to prevent the deterioration of hay at moisture contents up to 35 per cent, though for ammonia to be effective the hay to be treated must be placed inside a sealed plastic cover. These treatments may also increase the digestibility of the hay (perhaps allowing the crop to be cut at a slightly more mature stage) and, in the case of ammonia, may increase its 'crude protein' content. Further development of this work is awaited.

But haymaking still remains a weather-dependent operation, and it is essential to match the acreage of the crop cut at any one time, and the conditioning applied to the cut crop, to the best available weather predictions—so as to exploit any likely spell of fine weather to the full.

HIGH-TEMPERATURE DRYING

Undoubtedly the most efficient method of conserving green crops is by artificial drying with hot air. Total loss of dry-matter, from standing crop to dried product, can be as low as 3 per cent; further, because the crop can be cut for drying at a much less mature stage than is practicable for hay, the nutritive value of the dried greencrop can be much higher. Because of this potential a great deal of research was carried out on high-temperature drying ('grass-drying') during the 1960s, and the First Edition of this book, in 1972, described the considerable expansion of grass-drying that was then taking place. This expansion was to a large extent based on the results of research, much of which was carried out in close co-operation with farmer members of the Association of Greencrop Driers. Output of dried grass rose from about 65,000 tonnes in 1965 to 200,000 tonnes in 1972, and further expansion was predicted.

However, grass-drying requires the burning of fossil fuel, generally oil, to evaporate the water present in the fresh crop; up to 300 litres of oil are needed to produce one tonne of dried grass from a crop of 80 per cent MC. Thus the sharp increase in the price of oil during the 1970s made grass-drying very much more expensive, and considerably reduced the competitive position of dried grass as a livestock feed. As a result output has decreased markedly from the peak level reached in the early 1970s, to about 50,000 tonnes, and most of this is now sold in the premium market for poultry feed. The operators who have continued drying have remained competitive by wilting cut crops in the field before bringing them to the drying plant and thus greatly reducing their fuel consumption and raising their drier output; some have also installed dewatering equipment, which squeezes out some of the moisture from the crop before it is dried. Most of the existing drying plants were, however, installed in the 1960s and are not being replaced as they become obsolete, so that further decrease in the production of dried grass seems likely.

Perhaps, though, we should note several trends which, if they continue, might modify this pessimistic assessment of grass-drying. Firstly, while the selling price of dried grass per tonne used to be less than that of barley, it is now higher—reflecting the practical experience of the high nutritive value of the product which had been indicated in the earlier experimental work. The recent marked fall in the price of crude oil, if it is sustained, could also greatly reduce the cost of drying. Further, as continued restraint is placed on cereal-growing in the EEC, so as to reduce the present high cost of support, the search for alternative cash-crops on arable farms might encourage renewed interest in grass-drying. The problem is that this would require new investment in drying equipment, which seems unlikely; a plant to harvest and dry 2 tonnes of dried grass per hour would now cost well over £2 million. Thus, despite the undoubted technical and biological efficiency of grass-drying as a method of forage conservation, our forecast is that it will play at most only a minor role in future forage conservation, and it is not considered in detail in this book.

ENSILAGE

Quite distinct from crop conservation by drying is the process of ensilage, in which the wet crop is preserved by chemical action. As an early Ministry Bulletin, No. 37, noted, 'it is not necessary to be a biochemist to make good silage, but some understanding of the silage process is certainly helpful in developing a successful silage system on the farm'.

Much research and development has been carried out on silage since the method was first introduced into Britain a hundred years ago. Most of this has been concerned with the preservation process and with the reduction of losses, and only more recently has full attention been given to the importance of the feeding value of the product, and the effect that different methods of ensilage can have on feeding value. Consequently we find that some earlier methods, which appeared to give efficient preservation, are now not recommended because of the limited value of the product as animal feed.

When freshly cut grass is stacked in a heap it quickly begins to heat up as a result of chemical reactions taking place within the crop. In the early stages this is mainly due to the oxidation of some of the soluble carbohydrates ('sugars') present in the crop by oxygen in the air which is trapped within the crop. The heat produced speeds up the process further, partly because chemical reactions proceed faster at higher temperatures, but mainly because the warm air rising from the heap draws in fresh cold air, adding to the supply of oxygen, rather like the draught through a domestic fire. Thus the first essential is to stop air entering the heap of grass. This can be limited by consolidating the heap, compacting the grass so that movement of air is restricted. But undoubtedly the most effective method of controlling air movement and heating is by covering the surface of the heap with a plastic sheet. *This acts by preventing warm air escaping and drawing in fresh air*. This principle is an important feature of the sealed silo systems, described in Chapter 7.

However, by itself the grass in the heap is still completely unstable, and even in the absence of oxygen the huge numbers of moulds and bacteria, present naturally on the crop when it is cut, can rapidly multiply and decompose the grass into a putrefying and evil-smelling mass—a good example is a compost heap of garden lawn-mowings. To prevent this the activity of the undesirable micro-organisms, particularly the clostridia, must be stopped, either by sterilising the crop or, more generally, by making it acid—for these moulds and bacteria cannot grow under acid conditions.

In practice almost all silage systems depend on making the crop acid. The most obvious way of doing this is by adding acid, and this was the basis of the system developed in the 1930s by A. I. Virtanen in Finland, in which a mixture of sulphuric and hydrochloric acids (AIV acid) was added to the crop before it was put into the silo; other acids such as phosphoric acid have also been used. This method was at one time fairly widely used in Scandinavia, but it suffered from the very evident disadvantage that the operator had to handle corrosive acids—and the perhaps less obvious disadvantage that the resulting silage was not readily eaten by livestock.

Most silage-making, however, depends on the fact that grass carries a natural population of bacteria (the lactobacilli) which, in the absence of oxygen, can ferment sugars to produce lactic acid. These bacteria are relatively insensitive to the acid they produce in contrast to the 'undesirable' bacteria and moulds, which are largely inactivated as the grass becomes acid. This acidity is measured in terms of pH, an index widely used by chemists but confusing to the silage-maker because *a decrease in pH represents an increase in acidity*. Also pH is expressed on a logarithmic scale; thus to reduce the pH of grass from 6.8 (the very slightly acid reaction of fresh grass) to 5.8 needs only one-tenth of the amount of acid needed for a further decrease to pH 4.8. Thus an initial reduction in pH needs only a small amount of acid, and can be very rapid. But the rate of fall slows down as more and more acid is needed for each further 0.1 unit reduction in pH—and also because the lactobacilli do become less active at lower pH levels.

The level of pH measures the concentration of the acid in the water present in the silage. Unfortunately the wetter the crop the greater the concentration of acid—and the lower the pH—needed to inactivate the putrefying bacteria, and so the greater the amount of acid needed to give a stable preservation. For it is wet crops—leafy grass, well-fertilised with nitrogen; grass/clover mixtures; and autumn grass—that are the most likely to have a low concentration of the sugars which the lactobacilli need to ferment into lactic acid.

As already noted sugars are fermented to acid only if no oxygen is present—that is under 'anaerobic' conditions. This places even greater importance on preventing air getting into the stack, both to limit the loss by oxidation of the sugars needed to be fermented to acid and also to create the conditions under which the 'lactic' fermentation can occur. But even when anaerobic conditions have been ensured, by consolidation of the crop and by careful sealing, the wet crop still may not contain enough fermentable sugar to reduce the pH to the level required for stable preservation. The activity of the moulds and putrefying bacteria is then only slowed down; they continue to decompose protein; more seriously they start to decompose the lactic acid itself. As the lactic acid disappears the damaging process of secondary fermentation sets in, producing the evil-smelling silage disliked by both animals and farm workers (and of course their wives). If good silage is to be made from such difficult crops, other measures, in addition to efficient sealing, must be taken.

Wilting

Problem crops will not ferment properly because they are too wet and because they contain too little sugar. The first step then must be, whenever possible, to reduce the moisture content of the crop by wilting it in the field before it is brought to the silo; less sugar needs to be available for fermentation to preserve a crop of 25 per cent dry-matter content (stable at pH, say, 4.2) than the same crop at 18 per cent DM, which must be acidified to a pH below 4.0 before it will store safely. (It is worth noting here the established convention, that the condition of crops for haymaking is generally described in terms of moisture content [MC], but for silage in terms of dry-matter [DM] content. Thus a crop at 80 per cent MC cut for hay would be referred to as having 20 per cent DM content if it was to be ensiled.)

The process of field wilting is discussed in detail in Chapter 5. It has the further advantages that it reduces the weight of crop that has to be carted from the field and loaded into the silo, that it reduces the amount of liquid (effluent) likely to run out from the silo (of great importance in avoiding pollution of rivers), and that it may also increase the effective feeding value of the silage, because stock can eat more of wilted than of unwilted silage.

Wilting is of course essential if the crop is to be stored in a tower silo (page 138). Crop dry-matter levels greater than 30 per cent are then generally advised, so as to reduce the pressure on the walls and the effluent loss resulting from the weight of crop within the tower. At this moisture content only a moderate degree of acid fermentation is needed for the silage to reach a pH in the range 4.5–5.0, at which it stores well because of the airtight conditions within the tower.

When the crop is to be stored in a clamp or bunker silo some wilting before storage should also be carried out whenever it is practicable. But it may not always be practicable. Wet crops are often wet precisely because the weather is unfavourable for wilting; the rate of wilting early and late in the season is slow, even when the weather is fine; and, particularly on smaller farms, the wilting operation can complicate the harvesting process. Thus, to back up the desirable process of wilting the silage-maker may need other aids, including the use of chemical additives which make the ensilage process less dependent on the natural sugar and moisture levels in the crop.

Additives

For example, if ensilage is uncertain because of a shortage of sugar in the crop, then it seems sensible to add extra sugar. Addition of molasses, at about 9 kg per tonne of fresh crop, increasing to 15 kg if the crop is very wet or clovery, has been practised for many years with varying success, mainly because of the difficulty of mixing the molasses uniformly with the cut crop. Non-uniform mixing results in well-preserved silage interspersed with patches of poorly preserved (high pH) material; during prolonged storage the putrefaction from these patches can spread, so that much of the remaining silage can be ruined.

Some farmers now fix a simple container for molasses on to the tractor towing the forage harvester, from which molasses, delivered by gravity on to the swath just before it is picked up, is mixed with the crop in the harvesting process. The relatively large amount of molasses needed and its slow flow properties emphasise the importance of an efficient system of refilling the container (incidentally this has been presented as an innovation; in fact a very similar system was demonstrated to members of the British Grassland Society during a visit to Ulster in 1959, although its significance was not recognised at the time). Other farmers have used a simple tractor-mounted applicator to 'dribble' molasses on to the crop at intervals during filling the silo; but this is likely to give less uniform mixing with the crop than addition during the harvesting process.

As ensilage involves chemical changes, it should also be possible to improve the process by the addition of chemicals, and this possibility has been studied by many chemists for many years. The AIV process has already been noted; but most interest has been in *reinforcing* the natural fermentation process, rather than in replacing it by the use of strong acids.

Yet, despite an immense research effort, chemical additives had virtually no effect on practical silage-making until the mid-1960s—and for precisely the same reason that molasses had failed, that no effective method was available to mix the chemicals uniformly with the cut crop.

Plate 2.1 Applicator for formic acid. Acid from the two 25-litre plastic containers is fed by gravity directly into the chopping mechanism on the forage harvester.

Most of the research that had been done was with mini-silos in the laboratory, far removed from the practical farm problems of silage-making. Many will remember the enthusiasm with which sodium meta-bisulphite was introduced—and its failure in practice because it was spread by hand as the layers of grass were loaded into the silo.

This situation was transformed by the introduction in the late 1960s of applicators which feed the additive directly into the cutting mechanism of the forage harvester (Plate 2.1). As noted above a similar system had been developed for molasses, but had never been widely used. But with the combination of efficient chemicals and effective applicators the use of silage additives now became a practical proposition. With the back-up of a considerable R and D effort, additives were rapidly adopted on many farms, and their use was certainly a major factor contributing to increased silage-making in the 1970s (Figure 1.2). The most widely used additives to date have been based on formic acid, either alone or in mixture with other chemicals. This acid was studied in Germany as long ago as 1923, but its use did not become practical until 1965, when Naerland in Norway developed a simple applicator similar to that in Plate 2.1. Based on the Norwegian experience, formic acid was introduced in the UK in 1967 and, after detailed trials on research stations, husbandry farms and selected commercial farms, was marketed in 1969.

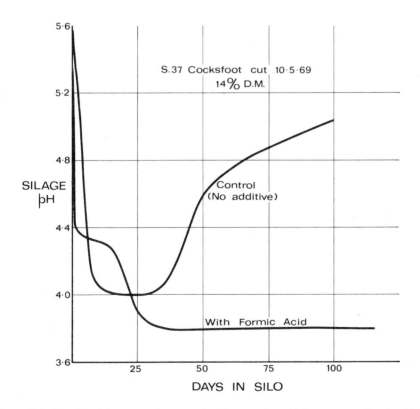

Figure 2.1   Formic acid prevents the secondary fermentation which can occur when 'problem' crops are ensiled

Results confirmed those in Norway, that an application rate of 1.7 kg of liquid additive, containing 80 per cent formic acid, per tonne of fresh crop, rising to 4.5 kg of additive with very wet or clovery crops, considerably improved silage fermentation. At these rates of application the pH of the cut crop quickly fell to below 5.0; this was not low enough to give safe storage but, as long as air was excluded, encouraged the development of a lactic-acid fermentation which then reduced the pH to the level needed for long-term storage.

In many cases the final pH of the silage was found to be no lower than would have occurred if the crop had been allowed to ferment without the additive. But the early natural fermentation of untreated crop tends to be slow, and some breakdown of the protein in the crop occurs; although the silage may finally reach a low pH, these breakdown products can cause a gradual increase in pH during prolonged storage, leading to deterioration of the silage. In contrast the rapid initial fall in pH when formic acid is added tends to prevent this protein breakdown. As a result, the silage remains stable over a long period (Figure 2.1).

Stimulated by the practical success of formic acid, and recognising that silage additives could offer a useful new market for chemicals, chemical manufacturers have subsequently introduced many alternative silage additives. Deciding whether to use an additive and, if so, which one, now poses an increasing problem for both farmers and advisers; for by 1986 more than sixty different additives were being marketed, not all of them backed up by adequate experimental or field-trial data. Yet to test this number of additives is clearly beyond the facilities of the UK advisory services, while the proposed comprehensive independent testing by universities, funded by manufacturers, now seems unlikely. In practice farmers must rely on summarised information on silage additives, which is produced, and updated at intervals, by ADAS. This groups commercial additives in terms of their main ingredients, and for each gives details of active ingredients, amount of active ingredients per kg of crop when the additive is applied at the recommended rate, and the treatment cost per tonne of crop. The following paragraphs, based on the ADAS classification, examine some of the factors that should be considered in deciding whether a silage additive is needed and, if so, which type is likely to be the most effective and economical.

(a) Inorganic acids
As noted, sulphuric and hydrochloric acids were used in the original AIV process; however, large amounts of acids were used so as to 'pickle' the forage and the resulting silages were not readily eaten by livestock. It was also difficult to mix the acids uniformly with the forage. Modern applicators allow these acids to be applied uniformly; smaller

amounts are also used than in the AIV process, so as to give a rapid initial acidification which encourages subsequent lactic-acid fermentation. However, inorganic acids are strongly corrosive, and great care must be taken when they are used, including protective clothing and goggles. But these acids, in particular in the form of 45 per cent sulphuric acid, are so much cheaper than any alternative additives that their use is rapidly increasing; they are already estimated to make up 80 per cent of the additives sold in Ireland. Some brands also contain trace elements, but it is doubtful if their extra cost is justified by any improvement in feeding value.

(b) Organic acids

Most commonly used is formic acid, whose action has already been described. Organic acids produce an immediate fall in the pH of the crop, but generally less than would be the case with inorganic acids, so that more acid needs to be produced in the subsequent fermentation to give a stable silage. Acetic, lactic and propionic acids have also been used but are rather less effective than formic acid; propionic acid, however, can help to reduce deterioration of the silage face when the silo is opened. These acids are less corrosive than inorganic acids, but care must still be taken when they are used.

(c) Mixtures of acids with formalin

In the early 1970s there was much interest in the possible use of formalin to preserve forage by sterilising it, as an alternative to acidification; it was also considered that formalin might improve feeding value by reducing the extent of protein breakdown in the silage (page 42). However, application rate in practice was found to be very critical, too much formalin damaging nutritive value, too little giving poor preservation. Thus formalin is now used only in mixtures, generally with sulphuric acid or formic acid, which give good preservation but require less formalin and acid than when each is used alone. Practical experience is that these mixtures are safer and more pleasant to use than the straight acids.

(d) Acid 'cocktails'

A number of additives contain one or more acids mixed with other chemicals which are claimed to improve their effectiveness. However, ADAS has found little evidence that such mixtures are effective unless the recommended rate of application contains the same amounts of the main active ingredients as when they are used alone. The recommended costs of treatment advised by different manufacturers are often very similar, indicating, at least in some cases, that additives are formulated to sell at a competitive price.

(e) Sugars

The use of molasses to reinforce the natural 'sugar' content in the cut crop has been noted. The effectiveness of molasses has been much improved by development of better applicators and by the use of formulations with lower viscosity which mix better. However, larger volumes have to be handled than in the case of other additives, and good organisation is needed to prevent this slowing down the harvesting operation. Molasses is completely safe to handle and use; it has the further advantage of adding some energy to the silage.

(f) Inoculants

Even when acids are applied, most silage-making depends on some fermentation of sugars in the crop to produce lactic acid. However, some crops may not contain enough bacteria, or the right types of bacteria, to ensure a proper fermentation, and there is now increasing interest in the use of additives (inoculants) containing bacteria to improve fermentation. Thus by 1986 some thirty-one silage inoculants were being marketed; all of these contained *Lactobacillus plantarum*, and most also contained other strains of bacteria. Unfortunately little experimental or practical evidence on the effectiveness of many of these inoculants has been available under United Kingdom conditions, as most development and testing had been carried out in the United States, where crops and conditions are very different. For in the United Kingdom the most common problem in silage-making is that the crop does not contain enough sugar; an inoculant may ensure that more of the available sugar is fermented to acid, but it cannot make up for an overall deficiency of sugar. For this reason several inoculants also contain enzymes which are claimed to break down complex carbohydrates in the crop, including cellulose and hemicellulose, to simple sugars which can then be fermented by the bacteria in the inoculant. Again, independent evidence is awaited on the effectiveness of these enzymes.

(g) Partial sterilants

The limitation of formalin for this purpose was noted in (c). A number of other additives, based on salts including sodium nitrite and calcium formate, can help to reduce initial bacterial breakdown in the silage while encouraging a longer-term acid fermentation. They have generally been applied as powders, and considerable care is needed to get uniform mixing. Some liquid additives now contain sodium acrylate, which has a similar action.

Practical experience to date has been that the additives in groups (*a*), (*b*) and (*c*) have proved the most effective, and they have certainly been the most widely used. However, it must be emphasised that additives are not essential to make good silage; their primary use must be to reduce the risks in making silage from problem crops—crops which are immature or very wet, particularly when wilting is difficult; crops with high protein or low sugar contents; crops with a high content of legumes; crops cut early or late in the season. These different aspects are brought together in Table 2.3; by assessing the different characters of the crop to be harvested this Table serves as a useful guide to whether an additive is needed, and if so the rate of application that should be used.

Certainly the first aim should always be to get some wilting in the field

Table 2.3 Guide to the use of silage additives

The number of stars (*) relating to the particular crop and harvesting conditions are added up. Above 20 * no additive should be needed; between 15 and 20 the normal rate of additive should be used; below 15 a higher rate of application is indicated.

	*****	****	***	**	*	*Number of stars*
Forage species	Italian ryegrass	Perennial ryegrass	Other grass species, or grass with clover	–	Mainly legumes	
Fertiliser N (kg N per hectare applied to crop)	–	–	Less than 50	50–100	More than 100	
Percentage dry matter content in crop	Over 25	–	20–25	–	Less than 20	
D-value of crop	–	Less than 60	60–65	Over 65	–	
Type of forage harvester	Precision chop	Double chop	Flail	Forage wagon	–	
Season	–	–	Spring and summer	–	Autumn	

(Data: AGRI and ICI Ltd)

before the crop is carted, if possible to between 20 and 25 per cent dry-matter content. A useful guide is that at 25 per cent DM some moisture can still be squeezed by hand from a sample of the chopped crop. But wilting may not always be practicable—or to wilt may mean delaying cutting the crop, or leaving the crop in the field for an extended period, both of which are likely to reduce its feeding value. Thus the use of an effective additive allows the crop to be cut when wilting is not possible, so that cutting and lifting can be carried out to a planned timetable, except under the most adverse weather conditions.

There is also now much practical experience of the advantage of applying a reduced level of additive to crops that have been over-wilted, say above 30 per cent DM, which may sometimes be difficult to avoid in hot weather. Such crops are difficult to consolidate properly to prevent air getting into the mass, and so readily overheat in the silo. An acid additive quickly inactivates the oxidising enzymes in the cut crop and the crop can then be consolidated and sealed without heating, with most of its sugar content still available to be fermented to acid.

Finally, as discussed in the following chapter, some additives appear to improve the nutritive value of the silage that is made, by increasing the amount of silage that livestock can eat, compared with additive-free silage made from the same crop. Such silage is likely to have a higher potential for animal production, particularly when the aim is for the silage to make up a high proportion of the total ration.

Whether hay or silage is being made, the feeding value of the final product is likely to depend at least as much on the 'quality' of the crop that is to be conserved as on the choice of the method of conservation. Some of the factors determining the nutritive value of conserved forages are now considered.

Chapter 3

THE FEEDING VALUE OF CONSERVED FORAGES

HAY AND SILAGE are seldom used as the only feed for productive stock. But because they can provide energy and protein on the farm more cheaply than most alternative feeds, the aim should be to include as much of them in the total ration as can be fed without reducing the level of animal production obtained.

This is not difficult with stock at low levels of production such as store cattle and sheep, rearing heifers, and late-lactation and dry cows. It is with the high-yielding cow, the beef calf, the finishing steer and the pregnant and lactating ewe that the problem arises of getting the animals to eat enough nutrients to meet their production requirements.

To get a high level of nutrient intake animals must be able to eat a lot of feed (high intake) and to digest the feed that they eat efficiently (high digestibility). In addition the feed must also contain adequate amounts of proteins and minerals for their particular requirements. Under most practical feeding conditions it is the amount of digestible dry matter that animals are able to eat—or more accurately the amount of digestible organic matter—which determines their level of production. Thus in order to make the most effective use of conserved forages in livestock feeding we need to understand the factors which determine (a) how much conserved forage animals are able to eat and (b) how efficiently they can digest this forage. Further, because conserved forages will generally be fed as part of a mixed ration it is also important to understand the way they are likely to interact with the other feeds which may be included in the ration, so as to make the best use of them under practical feeding conditions.

DIGESTIBILITY

Forage digestibility is most usefully described by the equation:

$$\text{Digestibility \%} = \frac{\text{Food digested}}{\text{Food eaten}} \times 100 = \frac{\text{Food eaten—dung excreted}}{\text{Food eaten}} \times 100.$$

That is, the lower the amount of dung excreted for each unit of food eaten the higher is the digestibility of that food.

Most experimental measurements of food digestibility have been made with sheep, by weighing the amounts of food eaten and of dung excreted over a balance period, generally of ten days. This, however, is an expensive procedure, impractical for the adviser or the farmer who wishes to assess the digestibility of a batch of hay or silage. Consequently, much research has been directed at developing laboratory methods which will estimate digestibility by analysis of a small sample of the feed. The most precise of these is the *in vitro* digestibility method, in which the feed sample is digested with liquor taken from a sheep's rumen, to simulate the process of rumen digestion. Because this is not well adapted to routine use simpler methods are now generally used, based on the fibre content of the feed (the more fibre the feed contains the less efficiently are animals able to digest it), or on the use of a naturally-occurring enzyme, cellulase, to estimate the digestibility of the cellulose in the feed.

Many earlier values were quoted in terms of dry-matter digestibility, based on the dry weights of feed eaten and of dung produced. But the feed may also contain mineral materials which, though digested, are of no energy value to the animal, as well as soil and sand which are not digested. Thus because digestibility mainly measures the energy value of the feed, the decision was taken in 1967 to use 'digestible organic matter', for it is only organic constituents of the feed which the animal can use for energy purposes. *The measure adopted was D-value, defined as the percentage of digestible organic matter in the dry matter of the feed.* From this it is possible, with reasonable accuracy, to calculate the Metabolisable Energy (ME) value of the feed, which is the measure of energy value now used in livestock rationing systems in the UK (page 32). But digestibility (D-value) is simpler to visualise in practice than ME, and is thus used in the following discussion.

The Digestibility of Forages at Cutting

The digestibility of hay and silage can be estimated in the laboratory; but of far more practical use is information on the digestibility of different crops while they are still growing in the field, which enables the farmer to predict pretty closely the digestibility of the hay or silage he is going to make.

It has long been known that, as the date of cutting of a crop is delayed, its feeding value falls. This is because the crop becomes less digestible as it becomes more mature and stemmy. Work at a number of centres during the 1960s, in particular at the NIAB at Cambridge, at the

Welsh Plant Breeding Station, at the West of Scotland College of Agriculture and at the Grassland Research Institute (now AGRI) at Hurley, showed that the digestibility of a growing grass crop can be predicted with considerable accuracy from a knowledge of the forage species in the crop and of its stage of maturity at harvest. For this purpose 'maturity' is defined by the number of days before the particular crop will reach 50 per cent ear-emergence (the stage at which ears have just emerged from half of the grass shoots that are going to produce ears)—or if cutting is later, the number of days after 50 per cent ear-emergence. This of course only applies to grasses; but as grass makes up the bulk of the forage crops cut in the UK the relationships between crop maturity and crop digestibility are widely applicable.

Results for the first spring growth of a typical grass variety, S 24 ryegrass, are shown in Figure 3.1. As the crop becomes more mature its digestibility, D-value, falls, initially quite slowly and then more rapidly after ear-emergence in mid-May (ear-emergence will be later in northern areas, see page 54). At the same time the yield of both dry matter and digestible organic matter in the crop increases. This means that, if a high yield of forage is cut, the digestibility of that forage is likely to be fairly low; conversely if high digestibility is required then this must result in a lower yield.

Clearly the decision made will depend on the class of stock to be fed, the other feeds available, and to some extent on the method of conservation to be used—young, highly-digestible grass is always likely to be more difficult to conserve than more mature, but somewhat less digestible, grass. But the important point is that, by using information such as that in Figure 3.1, a decision can be taken about the best stage of maturity (date) at which each particular crop should be cut for each specific feeding purpose.

Similar relationships have been developed showing the average yields and protein contents of a wide range of forage species and varieties at different levels of D-value. With this information, discussed further in Chapter 4, the farmer can decide the stage of maturity, and thus the date, at which to cut each crop so as to get the best combination of yield and feeding value for his particular feeding requirements.

The most accurate information on forage digestibility is undoubtedly that for the first growth of the year. Much of the national hay and silage crop is made from such first growths; but much is also made from regrowths, either from an earlier grazing, or from hay or silage aftermaths. It is more difficult to predict the digestibility of these regrowths, because this depends to a large extent on exactly when the earlier cutting or grazing took place.

A regrowth of S 24 which was grazed up to May 5th will tend to be

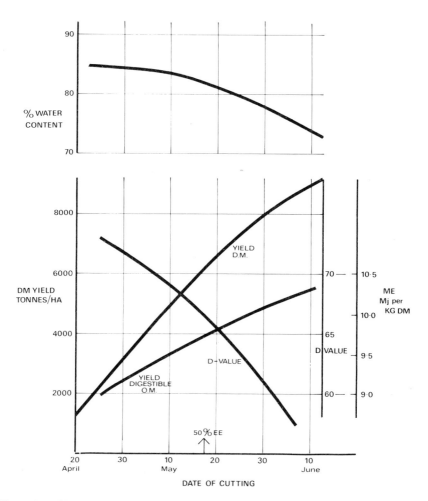

Figure 3.1   Changes in yield, digestibility and moisture content of S24 ryegrass during first growth in the spring. (Data: AGRI, Hurley)

stemmy, and its digestibility will fall rather rapidly; a regrowth from a silage cut taken on May 25th will contain relatively few seed-heads, and its digestibility will fall more slowly. Results published (AGRI, *Technical Report No. 26*; NIAB, *Technical Leaflet No. 2*), indicating the likely level of digestibility of regrowths of several grass varieties, following first harvests taken on a range of dates, can help in deciding the date at which each different regrowth should be cut to obtain forage of a particular digestibility.

Long leys and permanent pasture now make up a much higher proportion of the forage cut for conservation in the United Kingdom than in the 1960s, when a much bigger acreage was being sown to leys. This older grassland poses a problem because it is generally of more mixed botanical composition than leys. But, just as with leys, as permanent grass becomes more mature its digestibility also falls; because it often contains fairly late-flowering grass species, its digestibility is likely to be similar to that of the intermediate-flowering grasses cut on the same date (Figure 4.3), although the yield may be somewhat lower than that of a vigorously growing young ley.

The Digestibility of Conserved Forage

The importance of information on the digestibility of the growing crop is that it can be used to predict the digestibility of the hay or silage that will be made. In fact many experiments have shown that, with efficient conservation methods, the digestibility of hay or silage is very close indeed to that of the standing crop when it was cut.

But the proviso 'with efficient methods' is vital. The studies by Shepperson at NIAE, already quoted (Table 2.2), showed that barn-dried hay was very nearly as digestible as the cut crop, but digestibility and protein content were both much lower with field-made hay. This was particularly marked under unfavourable drying conditions, because of the mechanical losses of leaf and the leaching of soluble material by rain before the crop was fit to bale. This reduction in digestibility can be even more serious under farm conditions, where heating and moulding in the stack is all too common when hay is baled before it is fit.

With silage a decrease in digestibility can result from effluent running from the silo, and from overheating. But the main damage occurs when field wilting is attempted in unsuitable weather, and a considerable reduction in nutritive value can occur when crops are wilted for high dry-matter silage under unfavourable conditions.

Thus the digestibility of a conserved forage will nearly always be lower than that of the original grass, because of chemical changes and losses during the conservation process; the poorer the process the greater will be the fall in digestibility. The conservation methods described in this book aim to reduce this fall to an absolute minimum. But from a knowledge of the method used and of whether the losses, e.g. in field wilting, were average or above average, the corrections given in Table 3.1 allow a reasonable estimate to be made of the D-value of the hay or silage from the D-value of the cut crop, based on its species and date of cutting.

A more accurate estimate can, of course, be obtained by sending a sample of conserved forage for analysis by one of the state or commercial advisory laboratories. But such analysis can be carried out on only a few of the tens of thousands of different lots of forage which are conserved each year. With most lots of hay and silage an estimate of D-value, based on information such as that in Figures 3.1 and 4.3, and corrected for the method and efficiency of conservation (Table 3.1), should allow the feeds to be used effectively in the winter ration.

Table 3.1   Corrections to the estimated D-value of cut forage to allow for losses in the conservation process

Method	Subtract from D-value of forage
Barn-dried hay; good wilting	2
Barn-dried hay; moderate wilting	4
Rapid field hay; good weather	3
Rapid field hay; bad weather	5
Traditional hay; good weather	3
Traditional hay; bad weather	8
Low dry-matter silage; little effluent	0
Low dry-matter silage; some effluent	1
High dry-matter silage; good wilting	2
High dry-matter silage; moderate wilting	3
High-temperature dried grass; direct cut or good wilting	1
High-temperature dried grass; moderate wilting	2

D-values and Metabolisable Energy

In this book the feeding value of fresh and conserved forages is described mainly in terms of digestibility, D-value, partly because it is D-value that is generally measured experimentally, but also because the idea of 'digestibility' is much easier to grasp than any other measure of feed energy value.

However, ruminant feeding systems in this country are now based on Metabolisable Energy (ME), and for D-values to be used in working out rations they must first be converted to ME values. This conversion is based on considerable experimental evidence, with accuracy improving as more data become available.

Table 3.2 gives the estimated ME values for a range of D-values, which are now used by the nutrition chemists of the advisory services

Table 3.2   Estimation of the metabolisable energy (ME) value of hay and silage from the digestibility (D-value) of the feed

D-Value	Estimated ME (MJ per kg dry matter)
80	12.00
75	11.25
70	10.50
65	9.75
60	9.00
55	8.25
50	7.50
45	6.75

(Data: MAFF *Technical Bulletin 33*)

and the feeding-stuff industry. By using these data, together with tabulated data for other feeding-stuffs—compound feeds, cereals, sugar-beet pulp, etc.,—the production potential of a mixed ration can be calculated much more accurately than by the earlier Starch Equivalent system.

Feed Intake

The other main factor determining the productive value of a ration is the amount of it that livestock are able to eat. Under many practical feeding conditions the contribution of conserved forage to the total ration is limited by the small amount of forage that is eaten. As a result, large amounts of supplementary feeds may have to be given to bring the total nutrient intake up to the required level. If forages are to make a bigger contribution to livestock rations, this problem of low intake must be overcome.

First it must be noted that, although some feeds are 'unpalatable'—stock do not eat them because they dislike them—this is in practice seldom the main reason for low intake. The real problem arises with feeds which animals obviously find palatable, but where the amount they are able to eat is limited by some other factor—in particular, by low digestibility.

Observant stockmen have long known that animals can eat much more of a young, leafy hay than of an old, stemmy hay. The leafy hay is highly digestible, the stemmy hay is of low digestibility; and with a wide range of feeds we now know that, as feed digestibility decreases, the amount of feed that animals are able to eat also decreases. This means that nutrient intake, which is largely determined by intake × digestibility, falls rapidly as digestibility decreases. This lends added importance to the emphasis given to ensuring high D-value in the forage to be cut for conservation, to understanding the factors determining digestibility, and to the practical possibilities of predicting D-value from a knowledge of the species and stage of maturity of the crop being cut.

The relationship between intake and digestibility is most marked with fresh-cut (zero-grazed) crops and with crops conserved as hay. For example, young steers ate about 3.95 kg dry matter of well-made S 24 ryegrass hay of 70 per cent D-value, but only 3.40 kg of hay, made from the same species but cut four weeks later and of only 60 per cent D-value. The intake of digestible organic matter from the first hay was 2.76 kg ($3.95 \times \frac{70}{100}$) and from the mature hay 2.02 kg; in a feeding trial at Hurley the first hay gave 0.9 kg liveweight gain per day, but the late-cut hay only 0.7 kg, when both were offered as sole feed to yearling cattle.

Why do cattle and sheep eat less of low-digestibility feed? The reason is that the amount of feed that a ruminant can eat is mainly determined by the capacity (volume) of its digestive tract, and in particular of the large first-stomach, the rumen. It seems that an animal will eat until a certain level of 'fill' is reached within the rumen. The level of fill is rapidly reduced when a highly digestible feed is being eaten, partly because such feed is rapidly digested and broken down in the rumen, and partly because there is little indigestible residue left when it is digested; as a result the rumen empties rapidly, and the animal can then eat more food. In contrast, when the rumen is filled with feed of low digestibility it empties very slowly, because the feed is slowly digested, and because large amounts of residue are left. A low-digestibility feed occupies space in the stomach for a long time and this limits the amount of such feed that the animal is able to eat.

Differences in Intake between Forage Species

For some time it was thought that the relationship between intake and digestibility was so precise that it should be possible to predict how much of a given forage an animal would eat merely by knowing the digestibility of that forage; in other words, nutrient intake could be 'read off' directly from a graph such as Figure 3.1. However, it is now known that there can be considerable differences in intake between different forages, even though they are of the same level of digestibility. The most notable example is in the higher intake of legumes (white clover, red clover, lucerne and sainfoin) than of grasses of the same digestibility. Against this must be set the generally rather lower digestibility of the legumes than of the grasses at the stage of growth at which they are commonly harvested (see Figure 4.4)—though an exception is white clover, which is normally more digestible than the grass with which it grows in a mixed grass/clover sward (Figure 4.3). Stockmen, though, are well aware of the benefits from including some legume in hay mixtures. The problem arises in making hay from a legume-rich mixture without much of the legume leaf being lost. The need for the improved hay-making methods described later was never more evident than with grass/legume mixtures.

Even among the grasses we find differences in intake which could be large enough to be of practical importance. The most interesting is the generally higher level of intake of feeds made from Italian ryegrass than from perennial ryegrass of the same digestibility. The most surprising perhaps is the rather low intake of timothy, in view of the general opinion of the high 'palatability' of this species, particularly when made into hay. However, this observation could well be because much hay is made in mid-June, at which time timothy tends to be more digestible than the early varieties of Italian and perennial ryegrass which are mainly sown (Figure 4.3), and this could outweigh the theoretically higher intake of ryegrass than of timothy of the same digestibility.

The reasons for the higher intake of some forage species than of others may appear of rather academic interest; but their understanding could well lead to the breeding of new forage varieties of improved intake potential, and so make available to the farmer crops with higher potential for animal production.

The main reason for the differences in intake between forage species appears to be the same as that which relates intake to digestibility—that the more rapidly a food is digested and leaves the rumen, the quicker is space made available in the rumen and the more of that feed the animal can eat. We now know that when legumes and Italian ryegrass are eaten by ruminants they are more rapidly digested than, for example, perennial ryegrass of the same level of digestibility, so that more of them can be eaten. The differences in intake may be small (although cattle can eat some 40 per cent more dry matter as sainfoin than as timothy of the same D-value) and they are certainly less important than harvesting at the right stage of digestibility, or than improving the overall efficiency of the conservation system used. But once these aspects have been mastered it may well be worth considering the advantage of growing forage species which have an inherently high intake potential, so as to increase the level of nutrient intake when they are fed. Certainly, as already noted, this will be one of the advantages of having a proportion of legume in the forage that is to be conserved.

The Intake of Dried Grass

While dried grass seems unlikely to make a major contribution to future ruminant feeding in the United Kingdom (page 15), experiments with dried forages have provided much important information on voluntary food intake. When 'long' dried grass is fed, the relationship between intake and digestibility is similar to that found with hay. However, most dried forages are milled and pelleted in some form before they are fed, while even if the forage is not hammer-milled the process of packaging itself leads to a reduction in the particle size of the dried forage. As a result of this smaller particle size the voluntary intake of the pelleted forage is nearly always higher than that of the original fresh crop, or of the long dried forage before pelleting. This increase in intake, and so the benefit from pelleting, becomes more marked the lower the digestibility of the forage.

This effect is seen in Table 3.3, which records intake and performance data on young steers fed dried S 23 ryegrass, cut from the same sward at increasing stages of maturity—and so of decreasing D-value. As the date of cutting for drying was delayed the liveweight gains of the cattle fed the chopped dried grass were reduced, because both digestibility and intake decreased with increasing maturity of the crop. At each date of cutting part of the dried crop was also milled and pelleted before it was fed. Voluntary intake increased in all cases, and with it the resulting rate of gain, the greatest increase being found with the most mature feed. It is important to note though that, despite this, gains on the two more mature pelleted dried grass crops were lower than on the least mature feed fed in the chopped form. A similar improvement in feeding value results when hay is milled and pelleted before feeding; however, the economic gain is most unlikely to be enough to make this operation profitable.

Table 3.3 **Daily gains by four-month-old steers fed dried grass cut on three dates in spring, and either chopped or pelleted**

| | Date of cutting | | | | | |
	12 May		*3 June*		*28 June*	
D-value of grass	73		67		59	
	Chop	Pellet	Chop	Pellet	Chop	Pellet
Intake kg DM/day	3.72	4.00	3.00	3.72	2.58	3.17
Daily gain kg/day	1.00	1.18	0.77	0.95	0.45	0.63

(Data: Tayler & Lonsdale, GRI)

The Intake of Silage

The biggest practical problem involving low level of feed intake is found with silage. There is a great deal of evidence that ruminants may eat less dry matter as silage than as hay made from the same crop, and less also than of the fresh crop from which the silage was made. This was first demonstrated by Stanley Culpin at Drayton EHF, who recorded cattle gains of only 0.57 kg per day when silage was fed *ad lib* compared with over 0.68 kg when hay made from the same crop was fed. The problem is most serious with silage of low dry-matter content, and the intake and production potential of wilted silage are often higher than that of unwilted silage made from the same crop.

Much research during the 1960s attempted to identify the causes of the low intake of silage; high water content, acidity, toxic constituents, butyric acid, etc., were incriminated by various workers. But no single factor could explain all the observed cases of low intake. Then in 1970 Wilkins and his colleagues at Hurley found that low intake appeared to be associated with two main factors in silage. Firstly, the intake of high-moisture, well-preserved silage (pH less than 4.0) seems to be low because of the large quantity of acid which animals have to take in with each unit of silage dry matter eaten; for some reason there is a limit to the amount of acid that animals can eat. At the other extreme, the intake of wet, high pH (above 5.0) silage seems to be limited by its content of breakdown products from the proteins decomposed by the putrefying bacteria characteristic of such silage; this breakdown is indicated by a high content of ammonia in the silage.

In between these two extremes the intake of high-moisture silage in the pH range 4.0–5.0 can be low because of the combined effects of acidity and of some protein breakdown. Of the two, protein breakdown seems to have the more serious effect on intake and must be avoided; the revolting smell of putrefied high pH silage is also surely unacceptable under modern farming conditions.

Increasing Silage Intake

Although the mechanisms causing low intake are not fully understood, it does seem that to ensure high intake the silage-making process should aim to avoid both protein breakdown and excessive acid content. Yet high acid content (low pH) has always been the main target of the efficient ensilage process. This led to the search for methods of storing green crops at relatively high pH, but which at the same time would prevent protein breakdown.

The most commonly used method is to reduce the moisture content of the crop by wilting before it is ensiled; as noted on page 18 many crops of 25 per cent dry-matter content and above will produce a stable silage, under airtight conditions, even though the pH remains above 4.5, because less acid is needed to give preservation and prevent protein breakdown in the drier material. Both these factors have a favourable effect on intake; as a result the intake of wilted silage is likely to be higher than that of an unwilted silage made from the same crop which, even though well preserved, contains more acid.

A new approach was opened up by the observation that the amount of silage that animals will eat is increased when formic acid is used as a preservative. As Figure 2.1 shows, the pH of this silage is if anything lower than that of a control silage made without additive. But the formic acid also markedly reduces protein breakdown because of the rapid fall in pH during the early stages of the ensilage process and the prevention of secondary fermentation; this outweighs any effect on intake resulting from the higher acid content of the formic acid silage.

Production experiments carried out with dairy cows by Castle and Watson at the Hannah Research Institute confirmed the higher intake of silage treated with formic acid; results were particularly promising with red clover, a crop which is generally difficult to ensile without extended wilting—though even with formic acid some wilting is advisable to prevent effluent loss.

The discovery that certain additives can increase the voluntary intake of silage led to considerable research for new additives which preserve silage without the need for a high acid content (low pH). Yet despite some success under experimental conditions additives working on this principle, including formaldehyde and sodium acrylate (page 25) are little used. This is mainly because when they are used there is little margin for error under field conditions; in particular too low an application rate almost invariably results in poor preservation—and so low intake.

Thus the current advice to ensure high-intake silage is to wilt the cut crop before ensiling whenever weather conditions permit, aiming for about 25 per cent DM; to use an additive, preferably acid-based, whenever wilting is not possible; and even when wilting has been carried out, to apply a lower level of additive if the crop contains much legume, or if its natural sugar content is likely to be low, as in autumn cuts (Table 2.3).

Other Factors Limiting the Intake of Silage

It is also most important to minimise soil contamination of silage, as this can markedly reduce the amount of silage animals will eat, both through its indirect effect in giving a poorer fermentation, and its direct effect in making the silage less palatable. Any contamination with animal manure also makes silage unpalatable, and fields treated with cattle slurry should be given several weeks rest before they are cut for silage. There is also evidence that the intake of forages can be depressed by the presence on them of foliar diseases, particularly moulds and rusts. The depressive effect has been well documented with cereals, and grazing animals are seen to discriminate against diseased forage both at grazing and when the forage is conserved. In particular ensiling grass with a high level of fungal spores is likely to lead to poor fermentation, spoilage, and low intake; the extent of this problem in practice needs investigation.

CONSERVED FORAGES FED IN MIXED RATIONS

These studies of the factors determining the digestibility and intake of hay, dried grass and silage are an essential step in developing improved feeding systems. But only a step; for in practice most conserved forage will be fed in a mixed ration, and it is therefore important to know how far its 'basic' feed value may be modified by the other feeds used in the ration.

Balanced Rations

The feeding of farm stock is based on the following principles. First, that the nutrient requirements of an animal, in terms of ME, protein, minerals, etc., can be calculated from Tables of its requirements for maintenance and production, its current liveweight and its level of production of milk or liveweight gain. Secondly, that these requirements are met by the nutrient content of the total ration, calculated from Tables of the nutrient contents and the amounts of the different feeds eaten. In this way any deficiencies of particular nutrients, especially in the 'roughage' part of the ration, can be made up by cereals and concentrates rich in these nutrients.

This system worked well with rations based mainly on concentrate feeds, but problems arose when greater reliance was placed on home-grown forages during the 1939–45 war, and adjustments had to be made to the values for these feeds in the feeding tables. Partly this was because there was too little information available on the nutritive value of forages, partly because feed values are not always truly additive. Thus it is important to understand more fully the interactions between feeds if the most effective use is to be made of conserved forage.

Conserved Forages and Cereals

There is much evidence that a cereal supplement can reduce the digestibility (energy value) of the forage with which it is fed. Recent studies
suggest that this is because, when cereals are fed, the rumen contents
become more acid; that is, the pH in the rumen falls. Under these
conditions the micro-organisms in the rumen which digest fibre become
less active so that the fibre in the ration (which is mainly in the forage
component of the ration) is less completely digested, and the overall
digestibility of the forage is decreased.

But a more serious consequence is that the *rate* at which the fibre is
digested is also reduced. As a result, when a cereal-based supplement is
fed the rate of passage of the particles of forage through the rumen
falls—and so does the amount of forage the animal is able to eat. *This
means that a cereal supplement partly replaces, rather than fully supplements, the forage with which it is fed.*

The extent of this replacement varies with the type of forage being
fed, being greatest with hay and least with silage. Thus data from the
GRI indicate that the amount of hay an animal can eat may be reduced
by as much as 0.9 kg of DM for each 1.0 kg of DM of barley fed,
whereas the Hannah Dairy Research Institute has recorded an average
reduction of 0.5 kg of DM with silage. The reduction in intake of wilted
silage appears to be intermediate between that of hay and of unwilted
silage; thus the advantage of higher intake with wilted silage, compared
with unwilted silage, may not be fully realised when the silage is fed in a
mixed ration with cereals, as is likely in practice. Recent work also
suggests that the *extent* of forage replacement also increases with
increasing level of supplementary feeding; thus with dairy cows in early
lactation ADAS have found a replacement rate of only 0.2 kg of silage
DM when a low level of concentrates was fed, but that this increased to
0.9 kg at high concentrate feeding (in other words, for each extra 1.0 kg
of DM fed as concentrate the cows ate 0.9 kg DM less of silage).

The reduction in forage intake is perhaps not important when the
amount of hay or silage fed is being restricted; but it is serious when
high-quality forage is being fed *ad libitum*, with the aim of exploiting to
the full its high D-value and high intake potential. Thus ways of
minimising this undesirable effect of cereal-based supplements are being
sought. For example, feeding a given quantity of supplement in a series
of small feeds during the day, by using an out-of-parlour feeder, or by
mixing the forage and supplement in a 'complete diet' (page 168), should
limit the fall in rumen pH which occurs when large amounts of cereals
are fed at one time, as in parlour feeding. As predicted, this generally
leads to an increase in the quantity of forage eaten—though some

inconsistent results indicate that we still do not fully understand the regulation of feed intake.

Many cattle, in particular dairy cows, are now being fed a daily supplement of sodium bicarbonate to counteract the rumen acidity resulting from cereal feeding. Thus at Boxworth EHF cows fed 200–300 g per day of bicarbonate ate more hay and gave more milk than the control group (Table 3.4); milk fat percentage and butterfat production were also improved, almost certainly as a result of the more favourable rumen fermentation when bicarbonate was fed. Practical experience has also confirmed earlier experiments at Hurley that bicarbonate feeding

Table 3.4   Effect of including sodium bicarbonate in the concentrate ration of dairy cows fed on hay ad libitum

	Concentrate	Concentrate plus 200 g bicarbonate per day
Milk yield (kg per day)	22.9	24.3
Milk fat (%)	3.25	3.60
Concentrates fed (kg DM per day)	9.0	9.2
Hay intake (kg DM per day)	7.9	8.8

(Data: Boxworth EHF)

can increase the amount of low pH, high-moisture silage that dairy cows are able to eat, by reducing the effect on intake of the high acid content of the silage.

However, bicarbonate feeding has not proved effective in all cases; current advice is that it is most likely to be beneficial with early-lactation cows being fed a high level of concentrates; with cows on early spring grass; and with cows eating highly acid silage. Thus the individual feeder will need to test whether there is any response with his particular cows and feeds—and then to calculate whether the response will pay for the daily dose of bicarbonate.

When forages are supplemented with sugar-beet pulp or dried grass pellets, forage intake is depressed less than when the same weight of barley is fed, because these types of supplement do not reduce the pH in the rumen as much as when cereals are fed. Protein supplements, including oilseed meals and fish meal, also have less effect on forage intake, and the amount of silage eaten may even be increased; thus the Hannah has reported that silage intake may be increased by up to 0.1 kg of dry matter for each 1.0 kg of DM of groundnut cake or soyabean meal fed. As a result mixtures of cereals with proteins, as in conventional concentrates, are likely to cause less reduction in forage intake than straight cereals. In the case of silage more DM is eaten as the crude protein of the concentrate fed is increased—and with it the production potential of the total ration (Chapter 10).

The Protein Value of Conserved Forages

Part of the benefit from feeding a higher-protein concentrate arises, of course, because so much of the hay and silage currently made is of rather low protein content. Thus despite the considerable improvements in haymaking techniques of the last decade, the average crude protein (CP) content in hay samples analysed by ADAS in 1979, at 10.6 per cent, was only slightly higher than the 9.9 per cent of 1970–3; the CP content of silages was higher, at 14.2 per cent (reflecting the earlier cutting and higher protein content of forage crops made into silage), but this was very similar to the average figure ten years earlier.

Thus there is clearly scope, by earlier cutting and by greater use of legumes, to improve the protein content of the hay and silage made in the United Kingdom, so as to reduce the quantities of expensive protein supplements that need to be fed.

However, recent research suggests that high-protein forages, even when well preserved, may not be able to supply the full protein requirements of high-producing animals, in particular high-yielding dairy cows, and that some protein supplementation may still be needed.

This is because of the way forage proteins are digested in the rumen. These proteins are readily soluble, and when they enter the rumen they are quickly attacked and broken down by the micro-organisms there. One of the main products of this breakdown is ammonia; this can be used by the rumen organisms to produce the proteins needed for their own growth—and when this 'bacterial' protein passes out of the rumen into the hind-tract it is then digested and absorbed to provide an important part of the animal's requirements for protein (amino-acids). However, the protein in forages *can* be broken down so rapidly that ammonia is produced faster than it can be taken up and utilised by the micro-organisms. In that case the excess ammonia is absorbed directly from the rumen and excreted in the urine, and the feed protein from which it came is effectively wasted.

Forage proteins are thus described as 'rumen degradable'; in order that enough protein can be absorbed for the needs of high-producing animals they may need to be supplemented with 'non-degradable' protein—a source of feed protein that is only slightly broken down in the rumen, so that most of it passes on to the hind-tract, where it can usefully be absorbed. Again, supplementary proteins differ in their level of 'undegradability'; the most useful are fish meal, and oilseed meals that have been treated by heat or chemicals to make their proteins less readily available to attack by rumen organisms.

On the other hand some conserved forages, in particular maize silage, may not contain enough 'degradable' protein to provide the ammonia needed by the rumen organisms. Such forages can effectively be supplemented with a source of 'non-protein nitrogen', generally in the form of urea, which provides additional ammonia for the growth of the micro-organisms, which are subsequently digested and absorbed as amino-acids in the hind-tract; effectively urea, a cheap source of nitrogen, is converted into valuable 'feed' protein.

For their growth the rumen organisms also need an adequate supply of energy—hence urea is very useful as a supplement to high-energy (high D-value) feeds such as maize silage (page 117) and Italian ryegrass hay. But to be effective it is also important that the supply of urea is fed in phase with the supply of energy, so that the micro-organisms have energy and ammonia available to them at the same time. Thus urea is most efficiently used when it is mixed uniformly with the maize silage before feeding, or in a 'lick' mixed with salt to limit the amount eaten at any one time, from which the stock will take frequent small amounts during the day.

Fibre in Ruminant Rations

Ruminant animals are particularly adapted to digesting fibrous feeds; but must they have fibre in their diet? To answer this we need to distinguish between fibre as a chemical fraction and fibre as the structural component of 'fibrousness'. Too much chemical fibre reduces the digestibility of the feed, but a certain amount of 'long' fibre is needed to stimulate cudding, which is essential for effective funtioning of the ruminant digestive system, including preventing the rumen becoming too acid. This latter problem became particularly evident with 'barley beef' feeding. Barley feeding without any roughage, which had worked well under the strictly controlled conditions of the Rowett Research Institute, proved a hazard on some farms because of acidosis and bloat when cattle did not ruminate. This was prevented by feeding daily about ½ kg of hay or straw, which was enough to stimulate rumination.

In the case of the dairy cow there is a further need for long fibre in the diet, to ensure the correct rumen fermentation to give an adequate level of butterfat in the milk. Milk-fat synthesis needs a supply of acetic acid, which is one of the products of the fermentation process in the rumen. As the rumen contents become more acid following cereal feeding, less of this particular acid is produced—hence the problem of low butterfat levels in milk from high concentrate feeding. Conversely more acetic acid is produced on more fibrous diets, which help to maintain butterfat levels.

Most diets have in the past supplied plenty of fibre in the 'roughage' fed for 'maintenance'; however, the high levels of concentrates now fed to many high-yielding dairy cows may so limit the amount of forage eaten that enough acetate is not produced and butterfat level falls. Thus every effort must be made to ensure a proper rumen fermentation, particularly in early lactation. This can be assisted by feeding high D-value forage so as to reduce the level of concentrate feeding needed; by feeding the concentrate in a number of small lots during the day, so as to minimise its effect in making the rumen more acid; by feeding a supplement containing highly digestible fibre, such as sugar-beet pulp or upgraded straw (page 167). The residual alkali in straw treated with sodium hydroxide helps to maintain rumen pH; there may also be benefit from feeding a supplement of sodium bicarbonate (Table 3.4).

As nutrient requirements decrease later in the lactation it should be possible to feed more forage and less concentrates and there should be little risk of lack of fibre in the diet. In fact too often the problem is that the forage being fed contains too much fibre and so is of low digestibility and intake; more concentrates will continue to be needed than if a more digestible forage were being fed. With beef cattle too the commonest cause of low daily gains is that the forage fed is too fibrous and indigestible.

The Mineral Content of Conserved Forages

As with energy and protein, the requirements by different classes of stock for other nutrients, including minerals, are given in feeding Tables. There is also much information available on the mineral composition of the different classes of feeding-stuffs, including their contents of major minerals (calcium, phosphorus, magnesium, sodium) and minor minerals (iron, copper, zinc, manganese, etc.) However, these are *average* contents; the mineral contents of forages in particular are much more variable than their digestibility and protein levels, and it is difficult to predict the mineral content of a particular batch of forage.

Trends can be noted—the generally higher mineral content of legumes than of grasses; the high calcium to phosphorus ratio in legumes; the low contents of both sodium and magnesium in timothy—and these can indicate which supplementary minerals may be needed. But forage mineral content also depends much on the soil on which the particular crop is grown and the amount and type of fertiliser applied (for example, potassium fertilisers can reduce the sodium and magnesium contents in the forage grown). Thus it is a great advantage for the individual farmer to have a range of his crops analysed, so as to determine which minerals are likely to be deficient in the forages he harvests.

Of the major elements, phosphorus, magnesium and sodium are often found to be deficient, and problems with these elements, and with 'trace' elements such as copper, cobalt and selenium *could* become more common if greater reliance is placed on home-grown feeds, and less on bought-in compounds, which generally contain mineral supplements. Thus several silage additives are now advertised as containing trace elements, so as to make good these mineral deficiencies when the silage is fed. It would seem advisable to decide on such supplementary feeding in consultation with a nutrition adviser, who can take account of the mineral content of the forages grown on the particular farm, and of the other feeds likely to be fed.

This discussion of aspects of the nutritive value of conserved forages has inevitably included information which is repeated in later chapters on crops and feeding systems. To grow crops and apply feeding systems it is of course not essential to have this detailed information. In our view, however, some nutritional understanding will be increasingly important to the livestock farmer seeking to adopt the most profitable way to feed his animals.

Chapter 4

CROPS FOR CONSERVATION

MUCH OF THE HAY and silage made in the United Kingdom is cut from fields which are also grazed at some time during the year. In that case the particular crop species grown must also be suitable for grazing, especially where grazing is the main method of use. Cutting will have been introduced to utilise grass surplus to the requirements of the grazing stock and to maintain the pasture in good condition for grazing. In other cases the sward may be intended mainly for cutting, but with some grazing, generally later in the season; a few crops, for example forage maize, are grown for cutting only.

Basically most grass fields, whether cut or grazed, have a pattern of herbage growth throughout the year similar to Figure 4.1, with maximum growth rate during May and June, a check in mid-summer and then a lift in production during the early autumn. Clearly it is often difficult to match this to the daily feed requirements of many livestock enterprises. With some, for example the autumn-calving dairy herd and fattening cattle or lambs sold off during the summer, grazing requirements *do* decrease after mid-summer. But in most cases much more forage grows in the spring than can possibly be grazed by the available livestock; this 'surplus', often together with a smaller 'surplus' grown in the autumn, provides the main supply of forage to be conserved for winter feeding. Such conservation, generally as hay, has long been practised in traditional livestock farming. But the high losses in the systems used, and the poor value of the hay produced, made this operation very much of subsidiary importance to the grazing.

In contrast the newer methods of conservation, giving higher-value products with lower losses, have made forage conservation a much more important part of the overall operation, both in providing winter feed, and in improving the quality of the grazing by removing excess herbage before it becomes too mature.

Clearly, when the principal use is to be grazing, the quality of the forage cut will often have to take second place to the needs of the grazing system. However, in many regimes of grazing management,

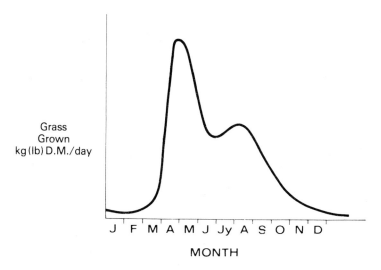

Figure 4.1 Daily dry matter production through the year from S 24 ryegrass, fertilised with N and irrigated

including rotational grazing and the so-called 3:2:1 system (Figure 4.2), cutting is fully integrated with the grazing so as to ensure adequate supply and high quality of both grazed and cut forage. Even under 'continuous' grazing, which has been adopted in place of more complicated grazing management systems on many farms, a part of the grazed area may be fenced off for cutting if grass growth is getting too far ahead of the stock.

Sown grasslands (leys) are particularly well adapted to a combination of cutting and grazing, and the information on yields and nutritive value, outlined in Chapter 2, can assist greatly in their management, and in deciding when leys should be cut for different purposes. Patterns of yield and D-value, similar to those shown for S 24 ryegrass in Figure 3.1, have been obtained for many other grasses and legumes. In practice there are now so many named varieties that it has been decided that a useful course is to group them into species and maturity types; thus the NIAB at Cambridge has found only relatively small differences in either yield or nutritive value between the different named varieties within any one maturity type of a species—reflecting the relatively small progress that has resulted from grass breeding (S 24, released in the mid-1930s, is still the standard against which other perennial ryegrass varieties are assessed), compared with, say, cereal breeding.

Thus Figure 4.3 shows the changes in D-value, with increasing maturity in the first growth, of the different types of several grass

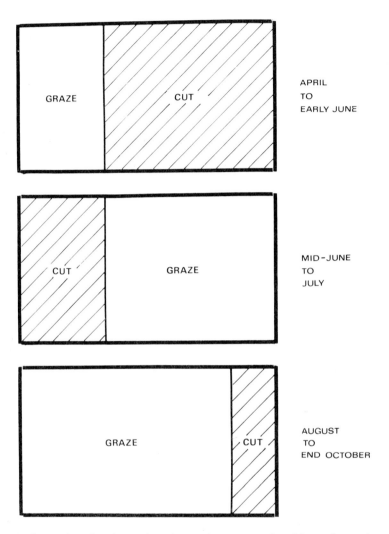

Figure 4.2   Integration of cutting and grazing to give a succession of forage for grazing

species, and of white clover, and Figure 4.4 shows the same for a number of legume species. From these figures several important points emerge.

- The D-value of all the grasses decreases as they become more mature, but with an indication that the *rate* of fall-off is less with later-maturing types.

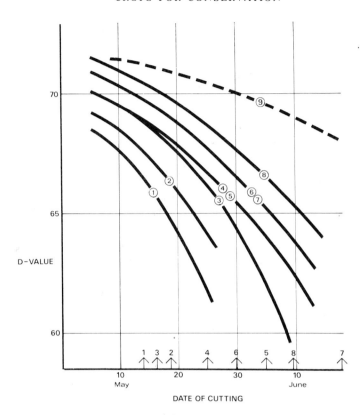

Figure 4.3 Changes in digestibility during first growth of different maturity types of several grass species, and of white clover. Key: 1 cocksfoot, early; 2 cocksfoot, late; 3 perennial ryegrass, early; 4 Italian ryegrass; 5 timothy, early; 6 perennial ryegrass, intermediate; 7 timothy, late; 8 perennial ryegrass, late; 9 white clover. Arrows indicate approximate dates of ear emergence. (Data: NIAB Technical Leaflet No. 2)

- On a given date different grass species, and different maturity types within a species, can differ considerably in digestibility. Thus all varieties of cocksfoot appear to be less digestible than the ryegrasses, while late-maturing varieties of perennial ryegrass (such as S 23) are more digestible than earlier-maturing varieties, such as S 24, when cut on the same date.

- White clover is always more digestible than the grasses; as a result mixed clover/grass swards are more digestible than pure grass swards. Red clover and sainfoin have digestibilities similar to S 24 ryegrass, but lucerne (3) is less digestible on a given date.

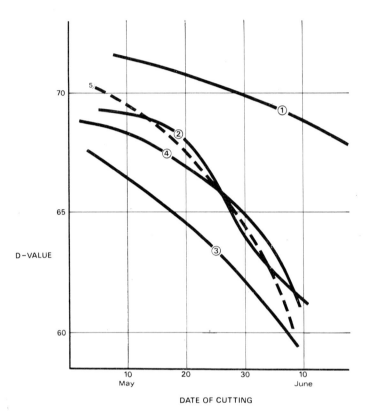

Figure 4.4   Changes in digestibility of first growths of several legume species. Key: 1 white clover; 2 red clover; 3 sainfoin; 4 lucerne; 5 early ryegrass

In making decisions upon which species and varieties to grow, yield as well as quality (D-value and protein content) must of course also be considered. An effective way of comparing crops is shown in Figure 4.5; this gives the yields of the different grass species and maturity types, noted in Figure 4.3, on the dates at which their digestibility falls to the same level (67 per cent D-value) during first growth in the spring. The high yields of the ryegrasses, followed by timothy, compared with cocksfoot are clearly seen. In addition to being higher yielding, the ryegrasses do not fall to 67 per cent D-value until the end of May or early June, when field conditions for cutting and wilting are likely to be much better than in mid-May, by which time the cocksfoots will already have fallen to 67 per cent D-value.

Figure 4.5 also shows the important point, noted above, that different maturity types within a species reach 67 per cent D-value at different

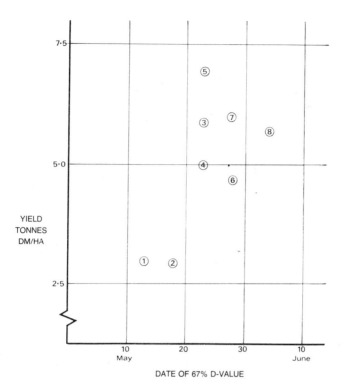

Figure 4.5   Dates at which the D-value of the first-growth of different grasses falls to 67 per cent and approximate DM yield on those dates. Key as Figure 4.3. (Data: NIAB Technical Leaflet No. 2)

dates. Thus early varieties of ryegrass reach this level around May 22nd, intermediate varieties about May 28th, whereas late varieties do not reach this level of digestibility until well into June. These differences may seem slight in relation to the full span of spring and summer, but they can be of real practical use in the management of fields for both cutting and grazing. For if different fields on the farm are sown to different maturity types of ryegrass a succession of first-growths of forage of very similar quality can be cut over a period of two weeks or more, with the latest varieties still being above 67 per cent D-value in early June.

The above discussion has mainly been concerned with the management of sown pastures. However, a major change in the United Kingdom grassland scene since 1972 has been the decreasing area that has been sown down each year to short-term leys, and the increased proportion of the total grassland under long-term leys and permanent grass. There are several reasons for this change: the relatively high cost of reseeding grassland; the reduced need for a grass 'break' on mainly-arable farms with the adoption of alternative ways of maintaining soil fertility and controlling weeds; and the recognition that permanent grassland, properly managed, can be nearly as productive as a newly sown ley.

A high proportion of the leys that are sown are found on dairy farms, and it is on these farms that different fields are most likely to be sown to different types of grasses, so as to give a better spread of forage production for both grazing and cutting. But the planned management of older grassland presents greater problems because their more diverse botanical composition makes their performance rather less predictable. However, as noted in the previous chapter, the relationship between the date of cutting and the D-value of much permanent grassland appears to be similar to that of the intermediate-maturity ryegrasses; this can usefully be checked on the individual farm by having several samples tested for D-value by the advisory service. But in general there is likely to be little spread in the dates at which different fields on the same farm reach a given level of D-value; thus a spread in cutting dates must be got by management.

A field heavily grazed until early May will have had a high proportion of the potential ears grazed off by the stock; thus the regrowth in early June will contain relatively few seed-heads, and will be of a higher D-value than first-growth forage from the same field if it had not been grazed. In fact its D-value is likely to be similar to that of the ungrazed field cut two weeks earlier—although the yield will be lower. Grazing of some fields on the farm will thus allow forage of fairly similar D-value to be cut over a period of two weeks or more. A similar management can of course also be used to get some spread of cutting dates from sown swards.

Such a spread of crop maturity, either by sowing fields to different grasses or by spring management, can make a major contribution to the overall efficiency of the conservation operation. Thus if all the cutting fields are of similar composition and are managed in the same way, all the forage grown will reach the same level of digestibility at the same time. To harvest all the forage at this level of digestibility would need a heavy investment in machinery and labour; alternatively, if equipment and labour are limiting, harvesting would have to be continued over several weeks, and the digestibility of the crop cut from the latest fields—and the digestibility of the hay and silage made from it—would be much lower than from the earliest fields. In contrast fields sown to different varieties, or grassland part of which is grazed up to early May, will provide forage for cutting over a range of dates, giving more efficient use of men and machines, and producing hay or silage of fairly uniform and predictable quality.

Fields managed in the same way are also all at risk to the weather at the same time. This is perhaps not too serious with clamp or bunker silage systems, which are not too sensitive to the weather. But with hay or tower silage wet weather at the time when the crop reaches the optimum stage of yield and digestibility means that the whole crop may have to be left, and to become less digestible, until the weather is fit for cutting. With a succession of crops the weather risk is at least reduced— though one can still expect the occasional period of two to three weeks of wet weather even in a British summer!

Finally, when conservation is being integrated with grazing, cutting a succession of swards produces a series of regrowths which become available in succession for grazing from mid-summer onwards—compared with the larger area of more uniform regrowth from a single grass variety cut at the optimum stage.

Clearly the use of different varieties should be kept as simple as possible. In most cases separate areas sown to two or three maturity types should be enough to bring about a real improvement in management for both conservation and grazing. This idea is of course not new; many farmers have sown fields to Italian ryegrass, which is ready for grazing or cutting earlier than intermediate perennial ryegrass. Certainly the greater the amount of forage to be conserved the more advantage there will be in ensuring a succession of forages for cutting.

Digestibility Levels

A comment is needed on digestibility levels. In this discussion, and in the Figures, a level of 67 per cent D-value has been used to examine aspects of differences in yield and digestibility between different forage species and maturity types. This is the level quoted in the NIAB *Technical Leaflet No. 2* as the stage of growth giving 'good quality', 'medium yield'. *But this is not meant to imply that 67 per cent D-value is the optimum level to aim at.* For, as noted earlier, the real use of information on D-values is to help the individual farmer to decide what to grow and when to harvest *for his own particular livestock enterprise.* This will need to take account of several factors:

● The class of stock to be fed; thus dairy and beef cattle need more digestible feeds than suckler cows and store cattle.

● The type of enterprise; the winter milking herd can justify higher-digestibility conserved feeds than the spring-calving herd. The latter will be at a low level of production during much of the winter, but will respond to a supply of top-quality feed in the weeks before and after calving, before grazing starts.

- The cost and availability of alternative feeds, in particular cereals and cereal-based concentrates.

- The stocking rate on the farm; at high stocking rates few acres may be available for cutting, and the strategy may be to take heavy cuts of relatively low-digestibility forage and to buy in supplements for winter feeding. To date this has applied in particular to smaller farms, on which more stock have been kept so as to increase 'the size of the business', and winter self-sufficiency has taken second place to a high stocking rate.

To assist the farmer in taking these decisions the farming press and radio now publish regularly during May and June estimates of the D-values of a range of forage varieties, based on samples cut and analysed by the advisory services during this critical period of the year. These data allow a correction to be made for the earliness or lateness of the particular season; for Figures 4.3, 4.4 and 4.5 are based on average data, whereas the date of 50 per cent ear-emergence, and so the date at which the digestibility of each type of grass falls to a particular level, can differ by over a week between an 'early' and a 'late' season. Date of ear-emergence, and so the fall-off in D-value, is also later the farther north in the country that the forage is being grown, maturity being up to two weeks later in northern Scotland than in the south of England; thus different D-value forecasts are also made for different regions of the United Kingdom. From this information the decision can then be taken as to how many days before, or more generally after, 50 per cent ear-emergence the forage should be cut to get material of the required digestibility, using, for instance, Figure 4.3 to indicate the rate at which D-value is likely to be changing.

The various factors to be considered in deciding the optimum D-value to be aimed for are considered further in Chapters 10 and 11, which deal with the feeding of different classes of livestock. But the method of conservation that is to be used may also influence decisions on the crop to be grown, and the stage of maturity at which it is to be harvested.

Crops for Hay

Even with facilities for barn-hay drying available, weather conditions in the United Kingdom are seldom suitable for cutting for hay before the end of May, and possibly rather later in the north. This means that most hay crops are likely to have reached 65 per cent D-value before they are cut. As a result well-made barn hay is generally in the range 60–65 per cent D-value—although this is a very acceptable level. Cutting of crops for field-made hay will generally be a week or so later. Thus there is advantage in using swards based on late-maturing varieties, or swards which have been grazed up until early May so as to remove many of the potential seed-heads. The yield of the latter swards, even if well fertilised immediately after grazing, will be lower than if they had not been grazed, perhaps not unwelcome because of the better chance of making good hay from the lighter crop.

Until the late 1950s half a million hectares were sown down each year to Italian ryegrass/red clover mixtures, to be cut for hay in June and July. These swards were often left down for two years, and ploughed for winter cereals in the second autumn. However, the area sown decreased greatly under the combined effects of cheap nitrogen fertilisers (reducing the need for clovers) and more intensive corn growing with fewer break crops, coupled with increasing problems from clover stem-rot and eelworm. With the breeding of more disease-resistant tetraploid varieties of red clover it was thought that the acreage of this crop, mainly for making into hay, would again begin to increase. This did not happen, partly at least because of the general decrease in the acreage of forage crops being sown, but mainly because of the greater reliability of production from ryegrass swards fertilised with nitrogen. Similarly the predicted increase in the amount of lucerne grown has not occurred, despite the availability of new varieties more resistant to *verticillium* wilt, because of the greater ease of management of all-grass swards.

By far the main tonnage of hay, however, is likely to come from older leys and permanent pastures of mixed botanical composition. The aim must be, by adopting the improved field haymaking techniques described in the following chapters, for these hay crops to be cut earlier, and at higher levels of D-value and protein content, than has been traditional, so as to make a medium-quality winter feed with the least risk of wastage and spoilage.

Crops for Silage

A much higher proportion of silage, in comparison with hay, is cut as an integral part of management on intensive beef and dairy farms. Much of the forage cut is thus from swards sown to Italian or perennial ryegrass, heavily fertilised with nitrogen, and containing little legume. As long as it is not cut at a very immature stage, the herbage from these fields is likely to have a fairly high sugar content and to be very suitable for silage-making—possibly aided by an additive if the herbage is wet and wilting conditions poor, or if it is cut late in the season when sugar content is likely to be lower (Table 2.3). By sowing different fields to early- and late-maturing varieties, and grazing some fields to delay the rate at which they mature, a succession of crops of similar D-value can be cut over a period of two to three weeks during May and June, spreading the weather risk and reducing the peak load on both men and machines.

On some farms the fields cut for silage are completely separated from those used for grazing, either because they are too far from the farm buildings to allow easy access for grazing, or because they are sown to a crop suitable only for cutting, such as forage maize. As with crops for hay, it had been expected that there would be an increase in the use of red clover and lucerne for silage-making; but well-fertilised grass has proved easier to manage, and yields have been so much more reliable, that the area sown to legumes for silage remains small—and is likely to continue so unless there is a big increase in the cost of nitrogen, relative to the returns from meat and milk production.

In the past a disadvantage of silage-making has been that it has not been a very convenient or efficient method for conserving the relatively small lots of forage, surplus to grazing needs, which become available at intervals during most grazing seasons. Generally these have been conserved as hay, often of quite good quality because these relatively light crops may be cut during good spells of weather in mid-summer. However, the new method of big-bale silage (page 130) now allows quite small amounts of grass to be conserved quickly and with high efficiency; on many farms this now offers an attractive alternative to haymaking, particularly for conserving surplus grass later in the season, when conditions become increasingly unfavourable for haymaking even in small lots.

Manuring Grass and Forage Crops

Manuring is a wide subject, covering both artificial fertilisers and animal manures, different grass and forage species, and different soil and climatic regions. Thus only a few aspects are considered here, mainly dealing with the use of nitrogen fertilisers, which play such a key role in forage production.

In the absence of fertiliser N or animal manure, grass and forage crops can only take up, even from the most fertile soils, enough N to yield about 3 tonnes of dry matter per hectare over the growing season. Yield responds markedly to fertiliser N, increasing by about 20 kg of DM for each kg of N applied, up to about 300 kg of N applied per hectare. At higher levels of application the yield response to N decreases and maximum yield is likely to be between 11 and 13 tonnes of DM per hectare at about 500 kg of N per ha. Rate of response and overall yield levels will be lower on poorer soils and in areas where summer rainfall is below 300 mm.

300 kg N per ha, though, is well above the general rate of application of N in the United Kingdom, so that there is clearly scope to grow more forage on many farms, with good yield response to extra N. However, there has been a tendency to restrict the amount of N applied to crops to be cut for conservation because of the practical problems that can arise in dealing with large crops, particularly if they are to be made into hay. But with modern harvesting equipment there should be few problems in making very large crops into silage, as long as extensive wilting—to much above 25 per cent DM—is not sought, so that no limit need be put on the amount of N used on crops to be ensiled. But making hay with large crops in unfavourable weather continues to pose problems, even with the latest techniques, and levels of N used should probably be lower than for silage.

There appears little advantage in applying N before the middle of March—maybe rather earlier in the south-west—to fields that are to be cut; if possible the dressing should be split, with a second application in April as growing conditions improve. From a combined application of 100 kg of N an intermediate-ryegrass type sward should yield about 6 tonnes of DM per ha by the end of May. This can be increased to 7 tonnes per ha by raising the N rate to 150 kg; alternatively the sward could then be cut a week earlier, to yield 5 tonnes of DM, but 2–3 units higher in D-value. However, while this higher rate of application is appropriate for a crop to be ensiled, 100 kg of N per ha is probably the most that should be used for a crop to be cut for hay early in June.

A further advantage of earlier cutting is that the rate of regrowth is likely to be more rapid, and response to further applied N higher, than from a later cut; this is important in ensuring that enough herbage is available for grazing, or for a second conservation cut, in late June or July. However, response to N is generally lower in mid-season than in the spring.

There is now renewed interest in the earlier application of N fertilisers, stimulated by the Dutch concept of T-sum 200 as an indication of cumulative soil temperature. However, the main aim of early N application is to encourage the growth of grass for 'early bite', so as to allow winter-housed stock to be put out to grazing as early as possible—perhaps indicating that neither the amount nor the quality of the forage conserved during the previous summer were satisfactory! But yield response to earlier-applied N is generally well below the 20 kg of DM per kg of N, noted above, particularly if a mild winter is followed by a late, wet spring. If the sward is allowed to grow on until the end of May the yield is likely to be lower than if the same quantity of N had been applied in March and April, rather than in February, or even earlier. Thus there is no advantage in very early application of N to swards which are to be cut for conservation. In practice a useful guide to the amount and timing of N applications is the 2 kg of N per day, developed by Fisons from their work at North Wyke. For example, if the farmer is planning for a cutting date of May 28th so as to get a crop of the required D-value, and intends to apply 120 kg of N per ha, then it is a simple matter to calculate that the fertiliser should be applied some sixty days previously, that is during the last week of March.

In contrast to the grasses, legume species are able to fix up to 200 kg of atmospheric N per ha through the activity of the bacteria contained in nodules on their roots. However, these bacteria appear to be less active when there is a good supply of available N in the soil; thus legumes growing in a grass/legume mixture well fertilised with N may not be able to compete effectively with the vigorously growing grass, so that their yield, and their contribution to the total N supply, may be reduced. This can be particularly serious with the more prostrate-growing white clovers, and there is likely to be little clover in a heavy conservation cut taken in May or June—or in the following regrowth. To maintain a reasonable clover contribution in a mixed sward (valuable in keeping up the digestibility of the sward after mid-summer) the spring application of N should be limited to 80 kg per ha, and the crop should be cut early so as to minimise the competition from the grass.

The more erect-growing legumes—red clover, lucerne and sainfoin—are more competitive, but can still be weakened if the grass component is heavily fertilised. On lighter soils there may be advantage in grazing these swards during the winter so as to reduce grass competition.

Forage Maize

Forage maize is not new; large crops of the variety White Horsetooth used to be grown, but harvesting was not well mechanised, and large amounts of effluent flowed from silos filled with this very late-maturing and low dry-matter variety. The position was transformed in the late 1960s with the breeding, mainly in France, of new, earlier-maturing varieties, which were normally ready for harvesting around the end of September. This 'new' crop was taken up enthusiastically by members of the Maize Development Association. This brought together farmers, advisers and research workers, keen to exploit the potential of the earlier varieties, and largely as a result of their efforts the area of forage maize grown in the United Kingdom increased to about 40,000 hectares. For several reasons the area then declined; fewer dairy cows were being kept in the mainly-arable areas of the country, which are the best adapted to growing maize; poor conditions for both growing and harvesting in 1975 and 1976 led to the crop being given up on some farms; and many growers concluded that forage maize, despite its high potential yield, was a more difficult crop to manage than intensively fertilised ryegrass.

However, some farmers continued to grow the crop very successfully, and consistently made first-class silage for their dairy and beef cattle. Thus the recent introduction, again from France, of several high-yielding but even earlier-maturing varieties means, we believe, that forage maize could now usefully be grown on many more livestock farms.

Maize has the particular advantage that the digestibility of the whole plant increases as the cobs begin to form, and then remains at a high level as the crop becomes more mature. This is in marked contrast to the grasses, whose D-value falls steadily after ear-emergence. At the same time the yield of the maize crop reaches its maximum value and its moisture content also falls (Figure 4.6). Against this, however, both weather and soil conditions are becoming less favourable for harvesting after the end of September in the south of the United Kingdom, and perhaps even earlier farther north. Hence the importance of the new earlier-maturing varieties, which will give a high yield, and should reach

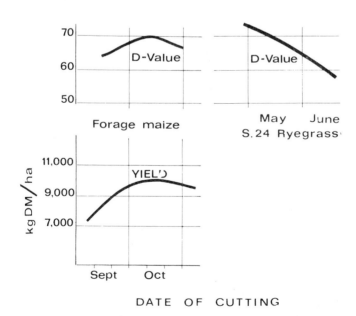

Figure 4.6   The yield and digestibility of forage maize cut on different dates (120 kg N per hectare applied). Also digestibility of first-growth S 24 ryegrass

the 25 per cent DM content needed for silage-making, before the end of September.

To achieve high and consistent yields by that date does however demand care and attention to detail in growing the crop. Exposed fields above 150 m altitude should be avoided, and to plant heavy land is asking for trouble in a wet autumn. But the crop does well on a wide range of medium loams and limestone soils. It needs no more than 120 kg of N per ha, which should yield up to 12.5 tonnes of DM. It responds particularly well to cattle slurry, which can provide much of this N; in fact the possibility of disposing of large quantities of slurry in the spring on land that is to be planted to maize is a practical and economic advantage offered by the crop. Sowing should be by precision drill set to plant 148,000 seeds per ha at 76 mm spacing in 760 mm rows; under average conditions this should give the optimum population of 110,000 plants per ha. With narrower drilling the lifters on the harvester may not be able to raise fallen plants, and forethought at drilling to provide a layout that suits the harvester operation will save a lot of time—such as the planting of a 14–18 metre headland *round* the field.

Rolling assists moisture conservation. Weeds used to be a serious problem, but can now be completely controlled by herbicides, either incorporated in the seedbed before drilling, or applied as a split dressing both in the seedbed and on the surface after drilling. Where couch is a problem it can be controlled by growing maize on the same land in successive years, for 4 kg of atrazine per ha, rather than the usual 2 kg, can then be used in the first year—a level which would damage most other succeeding crops. However, bindweed may still be a problem, and specific control of this weed may be necessary.

Time of sowing is also most important, for the longer the period of growth the higher the yield of grain in the cobs, which gives the crop its high feed value. If possible seed should be sown in the week following the date at which soil temperature reaches 10°C; this is seldom before April 25th, even on favoured fields. With adequate moisture following precision drilling to a uniform depth the whole crop should emerge at the same time. This is important, for it allows precautions against bird damage to be concentrated into the shortest possible period, with stringing and dawn gun-patrols starting just before the maize begins to emerge. Rabbits do not seem to be much of a problem.

Once safely established no further operations should be necessary until harvest. Leaf diseases have not been serious, but it is wise to avoid fields where there has been a high incidence of eelworm in cereals. Wireworm and leather-jacket infestations, likely after ploughing old pasture, should be identified before sowing, and the soil treated with an insecticide such as gamma-BHC.

Specialist drills and harvesting equipment (page 117) have a major role to play in the success of maize-growing, and the crop is thus very suitable for the well-equipped contractor, whose services can make maize a practical proposition even for the smaller farmer. The extended period over which the crop is at high D-value makes it well suited for contractor harvesting, and some contractors now offer an all-in service— provision of seed, drilling, herbicide spray, harvesting and haulage.

In our view the maize crop has a great potential for improving feed production on many livestock farms. On suitable land it also offers scope for introducing a 'double-crop' rotation in which a winter-green cereal such as rye is planted in mid-October and grazed, or cut for silage in April, in time for the land to be planted again to forage maize around May 1st.

Kale and Crop By-Products for Silage

Kale, sugar-beet tops and vegetable wastes can also be ensiled. But kale, in particular, is difficult to make into good silage, as it is cut in the autumn when it is of high moisture content and can give problems with effluent; basically there seems little point in ensiling this crop when it could instead be cut and fed directly. Potentially the largest by-product is the tops and crowns from the 200,000 hectares of sugar-beet grown in the United Kingdom. With modern harvesting equipment these can be collected relatively free from soil contamination. However, they are generally very wet, with dry-matter contents in the range 14–20 per cent, and give much effluent when ensiled. This problem can be minimised by mixing the tops with 15 per cent by weight of chopped straw before they are put into the silo, but the D-value of the silage will then only be in the mid-50s because of the low digestibility of the straw. Many arable farms now have little requirement for conserved forage because they keep few animals, while to ensile sugar-beet tops makes an extra demand on labour at a critical time. Thus there will need to be considerable changes in farming systems, and in the economics of live-stock feeding, before more use is made of beet tops.

Crops for Grass-Drying

On the relatively few farms which have continued to operate grass-driers the provision of a steady flow of high-quality green forage for drying, so as to keep expensive harvesting and drying equipment operating for as many days as possible, remains a key to economic success. In fact in the 1950s grass-drying had largely failed because crops for drying were only available over a short season and many units operated for less than eighty days in the year. The operators who have remained in business have learned to manage crops so as to keep drying for at least two hundred days.

Programmes to secure a succession of crops for drying have been greatly helped by the information on crop yields and digestibility now available. Profitable drying requires not only a long drying season (to reduce overhead and labour charges) but also a quality product which will command a high price. The quality objectives of high D-value and high protein content cannot always be combined in all the crops dried, but every crop should at least be either of high digestibility, such as Italian ryegrass in mid-May, or of high protein content, such as lucerne.

Most perennial crops have a basic growth pattern giving the highest yield in May and early June, with less growth in mid-summer and then a lift in production in autumn (Figure 4.1). Some spread of growth for cutting can be gained by using different species and varieties, as already described, but the basic pattern remains. Thus the drier operator looks for other crops that will complement this pattern. Examples are forage rye for early cutting—not an attractive crop because of its high moisture content—spring-sown grasses for cutting after mid-June, whole-crop cereals and field beans for cutting in July (crops of rather low D-value, but cheap to dry because their moisture contents are low), and forage maize for cutting in the autumn.

Clearly, to secure the steady flow of crop needed for drying demands a high level of field management, and the expertise needed is that of the arable farmer. The advantage of large-scale operation is also seen; for much the same level of skill is needed to plan a cropping programme for 80 hectares as for 800, so that the management cost of the latter is much less, per hectare, than of the smaller unit. While successful grass-drying will continue in mainly-grassland areas—with the particular advantage of short transport distance to the livestock that will eat the product—the main production is likely to continue to be from operations in which crops for drying are treated as alternative cash crops in arable farming enterprises.

As the importance of forage conservation in the overall farming enterprise increases, so will the attention given to the choice of crops to be grown. At the one extreme the livestock farmer on permanent grassland will be concerned mainly with conserving surplus grass which his stock do not graze in May and June; at the other the grass-drier may manage his crops to supply the drier over a long season. But both will need to give more attention to the stage of maturity at which they harvest their crops to produce feeds that will contribute more to winter livestock feeding.

Chapter 5

MOWING AND SWATH TREATMENT

MAKING EITHER HAY or wilted silage requires water to be removed from herbage in the field as quickly as possible with the minimum loss of dry matter. To achieve this, mowing, conditioning and any other treatments applied to the swath must be related to the type and yield of the crop, the expected weather conditions and the stage at which the cut crop is to be removed from the field. High crop yields make the problem more difficult, and some loss of dry matter is inevitable with any method which requires drying to be completed in the swath. As drying progresses the leaves become more brittle and more likely to break off when the crop is moved. These small fragments are difficult to recover so that even during fine weather, and using the best techniques, as much as 10 per cent of the crop dry matter is likely to be lost.

During rain soluble nutrients are leached from a cut crop and the digestibility of the remaining herbage steadily declines. The loss of soluble nutrients increases each time the crop is rewetted; in particular the effect of a modest rewetting from a nearly dry crop can be more severe than the effect of heavy rain on freshly mown grass. Most crops cut for hay are likely to be affected at some time by rainfall or heavy dew.

The aim therefore should be to exploit the principles of crop drying established in Chapter 2, first by physically treating the plant surfaces and then by managing the swath so as to ensure that the maximum amount of moisture is lost under the prevailing weather conditions.

Ideally some form of mechanical treatment should be applied at the time of mowing, or very shortly afterwards. As a general guide the swath should then be moved whenever the surface layers are appreciably drier than the remainder. The precise action required will vary, but generally the aim is to expose the wettest part of the swath to drying conditions near the surface, at the same time opening up the structure of the swath to encourage air movement and penetration of the sun's energy. The frequency or form of these treatments cannot be dogmatically laid down—such decisions must always be related to the weather.

Basic Requirements of Mowing and Conditioning Equipment

Mowing is a critical operation. The basic requirements are well established. The mower should cut the crop cleanly to the desired stubble height, and the swath which is formed should have a regular cross-section and be evenly distributed along the field, with an open, well 'set-up' structure to encourage the exchange of moisture-laden air. Swaths mown for silage should be in a form which can be picked up cleanly by the forage harvester and of even density to enable a high work-rate. Machines which have a pick-up in line with the towing tractor, such as large balers, require a swath with a particularly uniform cross-section.

To achieve a high output the mower should be capable of operating continuously and without blockage at a high forward speed in a wide range of crops, including crops which are laid or have a heavy and dense bottom growth. For reasonably efficient operation less than 25 per cent of the time should be taken up by turning and transit between fields, or by delays for adjustments. The design should prevent hard objects being thrown upwards at the tractor driver, although it is not always possible to avoid the ejection of missiles.

Clearly the mower must work at a rate, typically 1.5–2.0 hectares per hour, which will deal with the total area of grass required each day. To achieve this output mowers with a narrow width of cut, below 1.8 metres, must work at a high forward speed—up to 15 km per hour—which may not always be practicable because of poor ground conditions and consequent driver discomfort. Systems requiring high output should therefore use wider machines, between 2 and 3 metres wide, which can achieve the necessary work-rate at lower and more comfortable speeds. The swaths from these wider mowers may need to be gathered to about 1.6 m width to accommodate the tractor wheels and to match the subsequent handling and harvesting equipment, but the drying rate in the swath may then be significantly reduced. However, maximum outputs with forage harvesters are obtained when working in heavy swaths, so that if only a small amount of wilting is required a wide swath gathered in this way is to be preferred.

The power requirement of the mower must also be well within the working capacity of the tractor. This may seem self-evident; but it is prudent not to underestimate the power needed because rate of work, standard of mowing and recovery of crop dry matter are frequently disappointing when an under-powered tractor is used. Particular attention must be paid to this in hilly areas.

Mowers used alone do not impart any specific physical treatment to speed up the rate of drying. However, combined mower-conditioners are now available which both cut and condition the crop. Several additional points must be considered when selecting this type of machine. Thus power requirement is likely to be higher, typically 25 kW per metre working width, and this must be taken into account when matching the tractor if output and performance are not to suffer. The additional mechanism also increases the susceptibility of the machine to damage from stones and metal trash, and design features aimed at reducing this risk, coupled with ease of maintenance and repair, deserve special attention. Finally the suitability of the conditioning mechanism for the particular crops likely to be grown must be considered.

Treatment after Mowing

As explained earlier, mowing the crop removes the growing plant from its source of water and allows wilting to start. However, for haymaking the amount of water that has to be lost may be three or four times the final dried weight of the hay, and rapid action must be taken to get rid of this. The sap moisture which diffuses out of the individual plants cannot evaporate quickly unless the swath is disturbed to allow at least a drying wind, and preferably the sun's heat, to penetrate the swath so that it dries evenly. To achieve this the crop must usually be moved at frequent intervals, generally with a tedder.

Using a tedder immediately after mowing gives an early boost to drying rate. The same high drying rate can never be achieved if the swath is left flat and compacted, as it is with a reciprocating cutter bar or disc mower without a conditioning mechanism.

Whether or not some method of primary conditioning is used, tedding is still likely to be needed for secondary treatment. The ideal tedder is a multi-purpose machine, able to mix the swath, spread it thinly over the ground and then to collect it together in neat rows so that only a small surface is exposed to rain. When drying conditions improve the un-covered stubble and soil between the windrows dry out quickly and the hay can then be moved sideways on to this dry ground.

Tedding increases the rate of water loss from the swath as a whole and, correctly applied, does much to reduce the risk of patches of mould in hay. Tedders however do little to condition the individual plant parts or to speed up rate of moisture loss from the thicker stemmy parts. One method of doing this is to crush the fresh crop as it is mown—or at the latest within twenty minutes of mowing, since treatment of the cut crop after it has lost its turgidity is less effective. *Crushing* is carried out by passing the cut crop between plain or ridged solid rollers. This flattens but does not shorten the stems, and has only a minimal effect on the leaves, especially in lucerne and red clover. *Crimping* involves passing the crop through corrugated rollers, which have some crushing effect, but which also kink or bend the stems at 50–100 mm intervals and may bruise the leaf. There may be some shortening of the stems, while more severe damage to the leaves can be caused if the crimper bars have been 'burred' by stones.

Both crushers and crimpers are designed to 'set up' the swath so as to assist air movement, and in terms of improved drying rates and good dry-matter recovery these machines have much to recommend them. But they were developed mainly for use in lucerne and clover crops, and have never been as successful in heavy grass crops, which are less uniformly conditioned and tend to choke the machines.

The more severe treatment of laceration and bruising, which occurs during the cutting operation with mowing machines such as modified flail harvesters and flail mowers, has the effect of splitting and shredding both stems and leaves; this, coupled with some short chopping, considerably speeds up the rate of drying. However, with such treatments soluble nutrients are extruded to the surface of the individual plant parts, which become sticky and adhere to each other so that the swath tends to become more compact. A further disadvantage of this treatment is that even light rain can then wash away much of the soluble fraction. Rain can also compact a partly dried swath, reducing air movement; herbage left in this condition will soon deteriorate and eventually decompose.

Types of Mowing and Conditioning Equipment

Cutter-bar Mowers

Some requirements of the ideal machine have already been noted. The aim here is to consider how the equipment available matches the needs of different conservation systems.

Reciprocating cutter-bar mowers, once the most common mowing equipment, are now much less widely used because, if they are to cut cleanly, considerable time is needed each day on knife maintenance, particularly when working on stony ground—although this is a task that does not require much skill. The standard of mowing may be poor in wet and laid heavy crops, and this can lead to 'bunching' along the swath following a succession of blockages, and to heavy loss from uncut stubble. In contrast, in standing crops without heavy bottom growth, these mowers can leave an even stubble and, because the crop is only cut once, there is little or no fragmentation. Dry-matter loss directly attributable to mowing is therefore small.

Cutter-bar mowers are comparatively simple, lightweight, and cheap to produce, with a low power requirement at about 1.5 kW per metre width of cut. Forward speed is within the range 3–8 km per hour, so that with a 1.5 metre cutting width, overall output can vary from as little as 0.2 hectares per hour in heavy going up to 0.6 hectares per hour in a standing ley. Over the season output is likely to be only 50–60 per cent of these 'spot' rates of work.

An alternative is the more expensive *fingerless double-knife mower* in which the multiple scissors-like action gives cleaner cutting under difficult conditions. It is less susceptible to blockage than the finger-bar mower, though 'bunching' along the swath can still occur in heavy crops. It is also less liable to damage by stones, because the high knife-speed throws stones away from the cutting surfaces. A high forward speed, from 10–14 km per hour, is possible, with output up to 1.6 hectares per hour using 2–3 kW at the pto. A definite disadvantage of this type of mower is the high maintenance standard required. Knife sharpening takes a long time; it needs special equipment, and the skill to operate it, if high performance is to be obtained all the time. Against this, some 6–10 hectares can often be cut between sharpenings. To avoid delays some operators buy three sets of knives and have them sharpened by the dealer, but this adds appreciably to cost, while control of a critical operation is also taken out of the hands of the farmer.

Rotary Drum and Multiple Disc Mowers

Rotary drum and disc types of mower (Plates 5.1 and 5.2) are now widely accepted by farmers. They do not suffer from many of the limitations of cutter-bar mowers, though they are likely to use between four and eight times as much power. Typical pto power requirement ranges from 7 kW to 15 kW per metre width of cut, depending on type; this can be doubled with a machine which has badly worn or blunt blades, or is incorrectly adjusted for cutting angle or height. For optimum output, and to ensure that power is not a limiting factor, it is therefore advisable to use a tractor of 35 kW or over. Fortunately knife maintenance on these machines is simple, and up to 40 ha may be cut with one set of knives under favourable conditions; on flinty land, however, and with bad blade setting, blade change may be necessary every few hectares.

Instead of the conventional method of cutting by shearing between two surfaces these machines depend on the force of a number of freely-swinging knives, with a linear speed of about 80 metres per second, to slice through the plant stems. Given adequate power, rotary mowers work efficiently and without blockage in heavy and laid crops, although experience is needed to get them to operate well under all conditions.

Plate 5.1   Two-drum rotary mower. Cut crop passes between the two contra-rotating drums (Photograph Power Farming*).*

Plate 5.2   Rotary disc mower: cutter-bar arrangement. Each disc carries three free-swinging blades and the outer disc is fitted with a 'cone' to form the swath (Photograph Power Farming*).*

Particular attention must be paid to adjusting the height of cut and to the fore and aft setting and angle of the cutter unit; if this is done properly, 'scalping', accompanied by excessive blade wear, uneven stubble, and a 'mane' left between the cutting discs, can be avoided. With both types of machine fragmentation of crop and uncut stubble should be low.

However, double cutting of the crop as it leaves the rear of this type of machine can lead to loss of small pieces of herbage, especially in the later stages of haymaking. Care must also be taken to avoiding running on top of the cut swath with the tractor wheels, as this compresses the crop and can lead to uneven drying and poor picking up. This problem is less likely to occur with wider machines, with a working width of over 2 metres, if the swath is collected to accommodate the tractor wheels. However, on even wider machines, cutting 3 or more metres, there may be problems in following ground contours on uneven or undulating fields, leading to inefficient cutting; a tilt angle to allow the cutter unit to swing above or below the horizontal can improve cutting with these machines.

Rotary mowers can be operated at a forward speed up to 16 km per hour and continuous operation at 10–13 km per hour is practicable. A typical overall working rate is from 0.5 to 1.5 hectares per hour per metre width of cut, the wide range reflecting the effects of different machines, operators, crops and ground conditions. This method of cutting gives little crop conditioning, the thickest and wettest parts of plants being placed at the bottom of the swath and the more easily-dried parts at the top. The alignment of the plants also tends to form a dense mat compacted in a narrow width with a small surface area. These all restrict air movement and drying rate so that tedding or similar treatment is generally needed to get acceptable drying.

Combined Mowing and Conditioning Equipment

Where very rapid wilting is required there is much advantage in using equipment which cuts and conditions the crop at the same time. This conditioning can be an integral part of the cutting action, or can be applied by a separate conditioning mechanism.

Flail Mowers

Application of the flail principle to mowing for rapid swath wilting was first achieved on a large scale by the conversion of forage harvesters to flail haymakers. The most important alterations included reducing the rotor speed and removing the shear bars or other obstructions to reduce laceration of the crop, and fitting a delivery chute to direct the cut crop back on to the ground. Machines of 1.2 metres width, modified in this way, can also be used as a first stage in the production of wilted silage. The chopping and some laceration imparted to the crop leads to quick initial drying even in a compacted swath, while loss of the cut crop is minimised if the lifting action of a flail harvester is subsequently used to pick up the wilted crop. However, there can be unacceptably high losses when these machines are used to cut crops for hay because the smaller pieces dry rapidly, become brittle, and can be lost during baling.

A similar combination of mowing and primary conditioning is effectively carried out by the flail mower. Laceration without excessive chopping or leaf shatter, appropriate if the crop is to be made into hay, is achieved by operating the rotor at 800 rpm and using heavy-duty flails. Crops which are subsequently to be picked up for silage with a flail harvester can be cut with a higher rotor speed, up to 1,200 rpm, using double-edged swan-necked flails to chop the crop against a shear bar to get shorter chop-length. A tidy short stubble can be obtained without damaging the recovery regrowth of the cut plants providing the cutting height is set correctly (including adjustment of any devices fitted to avoid 'scalping' the crop), the flails are sharp, and both rotor speed and forward speed are matched to power input so as to prevent the flails swinging back over the uncut crop.

Flail mowers work particularly well in laid crops, although they can leave a straggly stubble if they are operated with the 'lay' of the crop. Freedom from blockage and 'bunching' is a feature of these machines as long as enough tractor power is available. Output, which depends greatly on crop type and power availability, varies from 0.4 to 1.2 hectares per hour for a 1.5 metre cut, and up to 1.5 hectares per hour with a 1.8 metre cut. Power consumption at the pto is from 12 to 24 kW per metre of cut, so that at least 35 kW must be available to maintain forward speeds up to 8 km per hour. On farms with a high-powered tractor an output of up to 0.6 hectares per hour for each metre width of cut should be possible under most conditions.

Knife maintenance and repairs are much less than with reciprocating cutter-bar mowers, because stones blunt rather than break flails. Loss of sharpness of course reduces cutting efficiency and increases power requirement, and there will often be need for a major overhaul between seasons.

Swaths which have been flail mown may occasionally need some secondary treatment before drying has been completed, especially after heavy rain has fallen on a crop that is nearly dry enough for baling. However, any following equipment must be chosen and operated with care; often a finger-wheel side-rake and turner will be all that is needed, but if a tedder is used it must be of a type that can be adjusted to give a gentle handling action.

Flail mowers are often heavy and cumbersome to use and their high power requirement can prove a limiting factor to output. The high losses of dry matter which *can* occur when they are used for haymaking means that they are not ideal for mowing and conditioning. Their use should be restricted to grass crops, or to grass-dominant mixtures, because their lacerating and chopping action can cause severe losses with legumes.

Mower-Conditioners

As an alternative to using the types of mowing equipment already described, the desirable features of different types of mowing and conditioning equipment can be combined into a composite unit.

The early versions of mower-conditioners used in Europe had been designed initially for use with lucerne in the United States, and were based on a reciprocating cutter-bar mower followed by a pair of crushing or crimping rollers. These suffered from the deficiencies of the cutter-bar mechanism, and did not prove successful with the heavy grass crops typical of the United Kingdom.

Equipment now developed for European conditions is based on either drum or disc mowers with a working width between 1.65 and 3.0 metres. Disc mowers have been fitted with intermeshing crushing rollers (Plate 5.3); these work effectively on an even flow of a medium-yielding crop, but are less effective on the thick swath from a heavy crop as only a small part of the crop comes into contact with the rollers. Drum mower-conditioners typically incorporate a rotor fitted with tines or beaters positioned at the rear of the cutting drums which penetrate the crop as it is concentrated at this point. Both types of machine leave a swath with a mixed and open structure which is particularly beneficial in the early stages of drying.

The ideal machine to combine mowing with mechanical conditioning would have the following features. Work-rate should at least match that of the equivalent 'plain' mower, but without a substantial increase in power requirement and with no greater risk of blockage or breakdown; the structural strength of the conditioned crop should be maintained so that the swath does not settle; the stems should be more severely treated than the relatively fragile leaves which dry more rapidly; any damage to the plant tissues should not increase susceptibility to leaching by rain or cause loss of small fragments; it would be advantageous if the severity of treatment could be varied according to the type and yield of crop; and

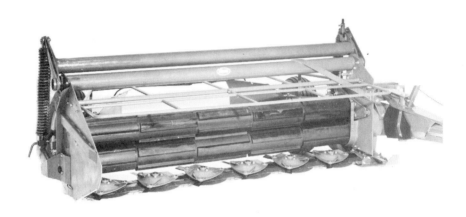

Plate 5.3   Disc mower-conditioner. Intermeshing crushing rollers are rubber faced to avoid fragmenting the crop and to leave a swath with a mixed but open structure.

the swath produced should retain an even and open structure, with a large proportion of the slower-drying stems exposed on the surface.

Whilst these principles may now seem obvious they only became clear from analysis of the many years of field experiments directed by the late Gordon Shepperson at the National Institute of Agricultural Engineering. From these a conditioning concept was developed at Silsoe which forms the basis of many of the mower-conditioners in use today.

This development of conditioning by surface abrasion minimises the barrier effect of the waxy cuticular layer on plant surfaces and so allows moisture to be evaporated more readily from the surface without destroying its structure or causing severe deep tissue damage. This effect is achieved by using a system of rotating spokes to overcome many of the problems of roller crushers and flail mowers. The basic configuration first developed is shown in Figure 5.1. A horizontal rotor carries radially-attached spokes arranged in 'V' formation and resiliently mounted for protection against impact from foreign objects (Plate 5.4). Operating at 850 rpm this rotor accelerates the crop as it leaves the cutting mechanism and projects it under a closely-fitting hood; the slip between the crop and the spokes abrades the waxy cuticle. Further, as the beaters penetrate into the crop, varying densities are theoretically treated more uniformly.

1   Forward extended safety curtain

2   Rotary mower drum

3   Bruising bar (Adjustable vertically & fore & aft)

4   Conditioning rotor

5   Rotor housing

6   Stripping/fluffing rotor

7   Lower swath forming deflector

8   Rear baffle

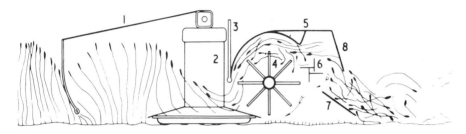

Figure 5.1   Method of operation of the NIAE drum mower-conditioner

Conditioning elements made from plastic with tough and resilient properties are now available (Plate 5.5). These offer several benefits, including lighter construction and less risk of damage to the following harvesting equipment or to livestock if an element breaks off. In a further development brush conditioners have been designed which retain the basic principle of the spoked rotor but with a larger number of conditioning elements. From the many configurations tested it has been concluded that a system of intermeshing tufted brushes is likely to be the most effective in terms of increased drying rate for a reasonable energy consumption (Plate 5.6). Laboratory studies have indicated that drying rate may be double that with the 'V' spoked rotor system. Machines based on this brush principle are now in commercial production. The dimensions and speed of the conditioning rotor are critically designed to maximise the conditioning effect while ensuring disengagement of the crop from the rotor to avoid wrapping and blocking.

In practice many factors interact to influence drying rate, notably crop type and yield, weather conditions and treatment following mowing. Extensive field-scale trials are still required to evaluate the relative performance of the different forms of conditioning mechanisms over a wide range of practical conditions; it is difficult at present to

Plate 5.4   Single 'V'-spoke element from a conditioning rotor. Each element is resiliently mounted to protect it against damage. Originally designed by the NIAE, the principle is used in many commercial mower-conditioners (Photograph NIAE).

Plate 5.5   Example of conditioning roller with plastic elements designed to achieve a similar effect to the steel 'V'-spoked rotor (Photograph NIAE).

Plate 5.6   Intermeshing brush rotor; each tuft consists of a number of plastic filaments (Photograph NIAE).

predict the exact effect of different mower-conditioners on drying rate in particular circumstances. However, by using such equipment it is likely that the drying time needed to reach 25 per cent dry-matter content (suitable for most silage systems) can be halved compared with earlier systems, while given good weather and effective post-mowing treatment a crucial two days may be saved with field-made hay. An important secondary benefit, particularly in silage-making, is that the mower-conditioner can present a regularly-shaped swath to the forage harvester without an intervening operation. However, the addition of conditioning rotors can double the power requirement compared with cutting alone, and most manufacturers now advise that tractor power of at least 30 kW per metre of working width should be available to maintain a good working rate under most field conditions.

Plate 5.7   Correctly operated, mower-conditioners will produce a uniform swath suitable for both haymaking and forage harvesting.

Operating Mowing and Conditioning Equipment

A number of general points apply to the operation of all types of equipment. Thus rolling grass fields early in the season is important; heavy rolling will reduce breakages and hold-ups caused by stones and other objects, and is especially important when flail and rotary mowers are to be used.

Effort made to produce a swath free from lumps at mowing will pay dividends throughout the whole haymaking operation (Plate 5.7). These heavy wet patches dry very slowly, and this either delays final baling or, if they are picked up before being fully dry, they form a nucleus for heating and moulding of the bales in store.

When working on soils which contain stones, particularly flints, all types of cutting machine should be adjusted so that damage to the blades is minimised. However, where close cutting can be carried out without damage to the cutting equipment, there is evidence from the Hannah Dairy Research Institute that the total annual yield from perennial grass swards is likely to be higher than with more lax cutting.

But unless the field is absolutely level both flail and disc mowers can 'scalp' parts of the sward, and this *does* reduce subsequent forage production, as well as contaminating the silage with soil. Thus under most practical conditions a cutting height above 40 mm from soil level is advised.

The importance has already been noted of applying the first conditioning or tedding treatment either at the same time or immediately after the crop has been mown; indeed with crimpers and crushers delay not only reduces effectiveness but may also make them difficult to operate. Swaths produced by modern mower-conditioners have an open structure which aids drying, so that the first tedding can be delayed until the surface layers are appreciably drier than the inside of the swath.

Secondary treatment following a severe primary treatment such as flail mowing, which causes heavy laceration and chopping, must be strictly graded according to the type of crop and the stage to which it has dried. Whenever possible the swath should be left as mown until it is ready to be windrowed for baling, using only a finger-wheel rake. If additional treatment does prove necessary this should be by inversion and turning with a finger-wheel so that the rolling action retains the small pieces of crop within the swath.

The amount of crop cut for hay at any one time should be matched to the available tedding, conditioning and baling capacity in such a way that each field, or part of a field, can be baled as soon as it reaches the required moisture content. This may at times demand almost continuous use of the tedder. Drying can be speeded up by collecting the half-dried hay into narrow compact rows overnight, and then spreading it out over the whole land area the following day as soon as the ground between the rows has dried.

Large dense windrows should be avoided while the cut crop is still very green. If rain does fall on to a windrow which has been prepared for baling, this should be split down into the original rows as quickly as possible, otherwise the rate of redrying will be very slow. Speed and efficiency of baling depend very much on the type of windrow presented to the pick-up reel. Maximum rate of baling is achieved when working in a heavy and dense windrow at a low forward speed, with five or six strokes of the baler ('wads') producing a 900 mm length bale. However, better-quality bales containing a larger number of wads, which may dry out and store better, can often be made by operating at a higher forward speed with a lighter windrow.

Effective operation of machines which make large round or rectangular bales depends greatly on the shape and form of the windrow. This should be about the same width as the baler pick-up, generally 1.5 metres, flat-topped and evenly distributed across the windrow. Uneven windrows, which have been formed by placing two swaths side-by-side but with a gap in between, can produce large bales with a badly-packed centre section which fall apart when handled. If the windrow varies in height from one side to the other round bales are likely to be produced which taper across their width—though this effect can be minimised by a skilled operator.

Swath-Handling Machinery

The ideal swath should not require any additional treatment between mowing and harvest but this is seldom possible in practice, except for silage crops for which little wilting is required. With almost all crops for hay it is necessary to tease open, mix and turn the swath as drying proceeds. When hot drying conditions prevail many argue that the crop should be spread in a thin layer over the whole field surface, with frequent turning; the care needed to avoid losing fragments of crop during the final stages of drying is evident. However, under the more likely conditions of dull weather the aim is to establish well-set-up swaths which intercept air movement across the field. In all cases the crop must eventually be gathered into a bulky regular windrow for baling.

For silage-making, where less extensive drying is required, swaths are more likely to be kept intact, especially if they have been cut with a mower-conditioner, and all that may be needed is to gather two or more swaths into a single windrow for efficient operation by the harvester. As already noted many farmers have recognised the advantage of using a wide mower-conditioner so that this extra operation, which also increases the risk of bunching and damage to the harvester from broken tines, can be avoided.

In most operations, though, swath-handling machinery will be needed, to perform any one of a variety of tasks—spreading, mixing, turning and windrowing. All these have similar basic requirements. Because the timing of all these operations is often critical, and occurs at a time of peak demand for both tractors and labour, a machine with a high work-rate but low power requirement is desirable; it should also have a wide working width, with good ground-following properties, to ensure that all the crop is moved with minimum risk of tine breakage. Plastic tines with tough and flexible characteristics are now increasingly used. In particular swath equipment used to collect crops for silage should be able to handle heavy swaths without bunching, at a rate well in excess of the capacity of the following harvester.

Many of the machines now used employ configurations of tines rotating in a near-horizontal plane, with the tines held out in their

Plate 5.8 Typical multi-purpose swath-handling machine with near-horizontal rotors. These machines can be used to spread, ted or windrow the crop, depending on the setting of the tines and the position of the crop collector (Photograph Power Farming*).*

working position either by centrifugal force or by cams (Plate 5.8). Some of these machines can perform only a single function, including machines with a single large-diameter rotor which operates with a side-delivery action, and others with two co-rotating drums arranged in staggered formation to deliver the crop to one side (Plate 5.9). If a large windrow is needed these side-delivery machines can be driven in alternate directions so that material from both sides is gathered into a single windrow. Machines designed only for spreading have a series of smaller, independently mounted rotors, arranged in line, up to 6.3 metres wide.

Other machines can perform several functions. Those with twin contra-rotating drums can spread, ted, or collect crop, depending on the position of a crop-guiding cage at the rear of the machine and the angle of the tines. With the cage raised the crop is spread, but when lowered it controls the flow of material and the machine can either ted a preformed row or collect a spread crop (Plate 5.10). More traditional tedders using a horizontal rotary drum are still in common use. When used for tedding these machines mix the crop effectively, while the wider versions, fitted

Plate 5.9   Two horizontal rotors in staggered formation give a gentle side-delivery action for forming windrows (Photograph Power Farming*).*

with crop guide doors, can form a windrow from crop collected across the width of the tedder.

A very wide range of swath-handling equipment is now available; the model's ability to meet the basic requirements noted above should be used as a guide in making a selection. In most cases correct setting and operation are critical. Except for the largest machines tedders are generally mounted on the tractor 3-point linkage. Single and double-row versions work at rates of 1.2–2.8 hectares per hour, with three- and four-row machines giving outputs up to 4 hectares per hour. A useful feature on some machines is the facility to adjust the tines to suit the crop being handled and the stage of drying, so reducing loss of brittle leaf.

Crop Loss and Drying Rate

Many factors affect the rate at which crops dry and the extent of dry matter lost during the process. Previous sections have indicated how machines may be selected and operated to maximise drying rate and

Plate 5.10   A multi-purpose swath-handling machine with the crop collector positioned to leave an open swath for rapid drying (Photograph Power Farming).

minimise crop loss. Detailed studies comparing different machines are limited, and depend very much on the specific conditions, particularly of the crop and of the prevailing weather. Recent data are also scarce. However, the results of experimental work over a number of years do indicate some general observations.

Losses in the form of uncut stubble are mainly of importance where a field is being used only for conservation, for in cases where the regrowth is grazed much of the crop left in the field will be eaten later by the grazing animals. Rotary mowers operated at the correct speed can cut as cleanly as the reciprocating mower, but if the knives are blunt or if they are operated at too high a forward speed in relation to tractor power, stubble loss can be high. On the other hand they can give lower stubble losses in laid crops or crops with heavy bottom growth, provided they have adequate power to operate at optimum forward and rotor speeds.

Loss of dry matter by fragmentation differs widely between machines. In an experiment in which grass was harvested from the field at between 55 per cent and 65 per cent moisture content, loss was only 2 per cent with a finger-bar mower and conditioner, 6 per cent with a flail mower, but up to 28 per cent when a modified flail harvester followed by a windrower was used. There can be even higher losses in cases where there is excessive chopping of the crop and the swath is handled aggressively. Other experiments have shown losses with flail and rotary mowers as much as 10 per cent more than with a cutter-bar followed by tedding.

Increasing the number of secondary treatments can speed up drying rate, but the advantage of this must be set against the risk that crop loss may be as high as 15–35 per cent of the dry matter if several treatments are applied to a leafy crop, in particular one that has been cut by a flail mower. If several treatments do have to be applied, in practice the most severe treatment should then be the first, for this can reduce the need for secondary treatment by at least one pass through the crop. Possible saving in field operations should always be taken into account when evaluating the advantages of alternative mowing and conditioning equipment.

It is difficult to generalise about drying rates; but certain patterns appear to be repeated under a wide range of crop and weather conditions. Thus the new designs of mower-conditioners are likely to reduce field exposure time by at least two days, compared with mowing and tedding. Tedding immediately after mowing gives an advantage of at least 48 hours compared with leaving the crop untouched in the swath.

A different situation pertains where grass is being mown and conditioned for wilted silage. In that case dry-matter loss is less of a problem because leaf does not become really brittle until its moisture content falls below 60 per cent, which is much drier than is needed for silage. Hence the aim is to get as much wilting as possible within 24 hours of cutting, indicating the advantage of more severe conditioning at cutting.

Regardless of the type and effectiveness of treatment, however, swath drying is still very dependent on weather conditions, and as the crop gets drier it becomes increasingly susceptible to damage from re-wetting, especially if it has been heavily conditioned. Losses can, however, be greatly reduced if the crop can be removed from the field before it is fully dry, and drying completed by fan ventilation, as discussed in the following chapter. Alternatively chemicals may be considered to allow baling to be done with hay at a slightly higher moisture content.

Non-Mechanical Conditioning

Various non-mechanical conditioning treatments, principally based on chemicals or heat, have been applied to the crop immediately prior to mowing, aimed at speeding up the rate at which it loses moisture after it is cut. These have all presented similar problems; in addition to the potentially high operating costs the treatments so far tested have tended to affect mainly the crop leaves, so exaggerating the difference in drying rate between leaves and stems which has always been one of the main problems in field drying. To be effective any treatment would have to be applied uniformly throughout the standing crop, and little progress has been made in developing effective practical systems.

Chapter 6

HAYMAKING

THE PREVIOUS CHAPTER examined methods of mowing and conditioning forage crops so as to encourage quick and uniform drying in the field. Once the required moisture content has been reached haymaking then becomes a materials-handling problem, with each subsequent movement of the crop adding to the cost without improving the value of the product.

BALES AND BALERS

The standard bale, 360 × 460 × 900 mm in size, and weighing between 14 and 23 kg, is a suitable unit for rationing from four to ten dairy cows. It is also convenient to manhandle and feed, especially when small quantities of hay are needed at some distance from the store. Baled hay needs only half the storage space of loose hay and its density and shape increase the capacity of transport vehicles. But standard bales are too small to allow individual mechanical handling, yet too large to be handled by augers or elevators in the same way as grain. Thus movement from baler to store and from store to animals nearly always requires hand work, and no method of handling gives a free-flowing system comparable with the mechanised systems available for silage.

Manual handling of bales can be reduced by handling them in unit loads by a tractor foreloader, either as groups of standard bales formed immediately behind the baler, or as larger bales approaching half a tonne in weight. Hay can of course be handled by methods other than baling; these generally involve further drying after removal from the field, and will be discussed separately.

Standard Balers

A very large proportion of the hay made in the United Kingdom continues to be handled as standard bales; though many farmers are turning to other forms of handling and conservation as the opportunity arises, many retain pick-up balers for handling at least part of their crop. Thus it is worth considering balers and baling in some detail.

Most balers at present used are of the slicing-ram type, in which a reciprocating ram, which presses the bale into the chamber, is fitted with a knife to cut through the flow of crop at the end of each stroke. Provided the knife is maintained in good order and registers correctly with the shear plate on the chamber, neat bales, in which the 'slices' are cleanly separated, will be formed. Bale density, although varying considerably for any one setting depending on the type and moisture content of the crop, can be altered over a very wide range, from 80 to 220 kg/m^3, by adjusting the spring-loaded pressure plates. Wedges may also be fitted in the chamber to increase resistance when the crop is dry. On most balers the bale length can be varied over a range from one and a half to two times the bale width, using a simple star wheel and tripping arm driven by the flow of crop through the chamber.

In contrast the swinging ram or press-type baler has a folding action, in which the successive charges are not positively cut. The bales produced are less tidy, and generally of lower density, than bales made from the same crop with a slicing-ram baler, but they have the advantage that air can penetrate more easily and so allow better ventilation, which can be an advantage with slightly-moist hay. However, few of these machines are now used in the United Kingdom.

There are many minor variations between makes of baler, but the ideal requirements are common to all. Thus the pick-up should be wide enough to deal with a heavy crop, and preferably able to take a swath of over 1.5 metres width. A combination of narrowly spaced pick-up tines, a freely-floating pick-up and a spring-loaded crop guide is the most likely to pick the crop up cleanly. The crop must then pass in a clean unobstructed flow from pick-up to bale chamber. Efficient packers, which can be easily adjusted, are essential if well-shaped bales are to be formed. If the crop is fed either too far or not far enough into the bale chamber the bales will curve away from the too-dense side; apart from the resulting nuisance of bales with strings becoming detached, such irregularly-shaped bales reduce the output of most handling systems.

There has been a progressive increase in baler ram speed, up to 80–90 strokes per minute at standard pto speed, and this has led to smoother running, especially when the ram is mounted on sealed ball-bearing rollers. Clearly this increased speed is an important factor in getting high output, but the uninterrupted operation is also assisted by having a heavy flywheel, which smooths out surges and reduces the peak power requirement caused by variation in swath density.

The baler knotters should be able to operate with a range of twines although minor adjustments, for example to twine tension, may be needed.

Wherever possible any bale-grouping device that is towed behind the baler should not subject the bale chamber to uneven loading. A rapid and easy method of converting the baler from the 'transport' to the 'working' position without the tractor driver leaving his seat is also a great advantage. For high performance the baler must have comprehensive out-of-season maintenance as well as daily attention when operating, particularly to lubrication of the knotter and needle assemblies.

The technique of operating a baler to produce well-shaped regular bales of reasonably even weight can only be learned in the field. Most of the *technical* data required will be included in a good instruction book, but a number of general points are worth noting.

Bale Density

The baler must be adjusted at intervals according to both the type and the moisture content of the hay being baled; otherwise bales will vary from very heavy in crops of moist leafy hay to very light in stemmy open crops of very dry hay—with bursting of strings at the one extreme and collapse of bales within the strings at the other. Adjustment will be needed at sunset, whether or not dew starts to fall, and there can be particular problems on the headlands of fields near hedges and woodlands.

Hay is normally baled when its moisture content is between 20 and 35 per cent. At the most likely level of 25–30 per cent MC bale density should be about 160 kg/m³, giving a standard bale weight of about 23 kg. With drier hay density will fall to about 110 kg/m³ and bale weight to 16 kg, while with hay at over 30 per cent MC density can be over 210 kg/m³ and bale weight 32 kg. If an extended chute is connected to the rear of the bale chamber to deliver the bales into a trailer, this will also increase bale density and weight.

If the hay is to be barn-dried it will probably be baled at a moisture content above 30 per cent, and bale density will then be very high unless the tension is slackened; a good guide is that it should be possible to thrust half a hand into the side of each bale after it has been released from the chamber. Wet bales can often only be kept down to a reasonable weight, say under 32 kg, by reducing the bale length to 700–800 mm. This shortening has the further advantage that the bales are then more likely to remain intact within their strings after they are dried.

Mention has been made of the number of wads (strokes) per bale, which is directly related to the rate at which the crop is being fed in and the ram speed. Bales with up to 20 wads, made from a light crop at high forward speed, dry out well if left stacked in the field but can become rather loose. For barn-drying the ideal bale contains 10 to 14 wads, whilst well-formed bales made from a large windrow at high forward speed will contain 8–10 wads.

Baler Output
Baler output varies widely according to the crop and the weather conditions, but general guidance may aid decisions on the quantity of hay which can be made at any one time and on the capacity of the ancillary equipment that will be needed to move the baled hay from field to store. Very high rates of work, up to 15 tonnes of hay per hour, have been recorded over short periods of operation using balers with high ram speed and an enlarged bale chamber, 460 × 410 mm, as compared with the normal 460 × 360 mm. But the output of most types of baler varies from as low as 3½ up to 10 tonnes per hour, with a common on-farm rate being 6–7½ tonnes per hour, equivalent to the hay from about one hectare. While this rate, of 250–300 bales per hour, is well below the potential of most balers, it is far in advance of most bale-handling systems. Without mechanisation a sizeable gang of men, working at 45 bales per man-hour, would be needed to clear bales being produced at this rate.

Handling Standard Bales

The mechanisation of bale handling is done in five stages: grouping, loading on to transport, transport, unloading from transport, loading into store.

Ideally bales should be taken directly from the baler to the store in a system in which each stage matches the output of the next. But in practice this is rarely feasible. Thus in most systems bales are generally formed into small stacks for a holding period, either in the field or under temporary cover. In this position the hay is protected to some extent from the weather while any heat produced from continued respiration, together with some moisture, is fairly effectively dissipated. This intermediate holding stage between baling and final storage also allows handling to be done in distinct stages—of advantage when only a small team of workers is available.

The shape and size of these heaps of bales should be planned as an integral part of the handling system; at the same time the formation of the groups of bales should not be directly linked to the baling operation if this in any way reduces baler output. The overall dimensions of any handling equipment must also be related to field size and topography and to building size and layout. In particular poor accessibility around buildings can limit the choice of equipment, and with handling systems based on large unit loads effective storage space can be reduced if the load dimensions are badly matched to the size of the building, leaving space too small to take further bale units.

The first priority in choosing equipment is to decide how many tonnes (bales) of hay will need to be handled in a specified time in order to avoid deterioration between baling and storage; for the feeding value of the final hay can be seriously reduced during this time. Timeliness is therefore a primary consideration. But clearly the rate of handling must also be related to the available labour force and to the amount of capital which can be invested in trailers and specialised equipment.

Handling Single Bales
Single bales can be lifted directly from the row and loaded to trailers with an automatic pick-up elevator at up to 180 bales per man-hour; with a tractor-mounted hydraulically-operated swinging arm bales can be tumble-loaded at over 400 per hour. With these methods a team of three men can haul and store up to 200 bales an hour, which at 70 bales per man-hour is double the working rate of most hand-loading systems.

The use of extended chutes and throwers, attached to the rear of the bale chamber, which convey the bales directly into the trailer, allows rapid clearing of the field, with the supply of trailers often being the limiting factor. However, stacking the bales on the trailer can be hard work, and if this slows down then baling must also slow down. Even with tumble-loading from the baler time can be lost in making turns in the field with this lengthy equipment, and in hitching and unhitching the trailers. As a result overall baling rate may be no more than 200 bales per hour.

Special-purpose bale loading wagons can be used as a one-man operation to pick up single bales and transport them. The wagon forms these into a block of bales which can then be deposited as a stack, or unloaded individually for hand stacking. The latter technique is likely to be preferred for hay as the stacks, once dropped, cannot easily be moved again. Another self-loading wagon consists of a spiralled conveyor system to pick up, transport and discharge individual bales. One of these wagons can collect a load of 88 to 130 bales in 10–20 minutes; however, it is specialised equipment and has not become established on farms in the United Kingdom.

Handling Bales in Groups
Single bales as left by the baler are easily damaged by heavy rainfall, and they also interfere with regrowth of the crop if they are left on the field for more than a few days. For these reasons, and also because of the advantage in mechanical handling of larger units, it is now common

practice to collect bales into groups. In many cases these are first formed into small intermediate stacks arranged either in rows down the field or on the headlands. As previously noted this can assist the organisation of the handling system by breaking it into a series of separate stages; at the same time some curing of the hay can take place in these small stacks. With the tops covered by plastic sheeting the stacks can be safely left on the field for some time, though the sheets should be moved at intervals to prevent condensation. This practice is much more common in the difficult climate of the north of England and in Scotland than elsewhere in the United Kingdom.

The first requirement is to collect the bales into groups after baling, but without interfering with baler output. Various forms of collector, towed behind the baler, have been developed for this purpose. With the manned sledge, baler output can be reduced by as much as 20 per cent because the heavy sledge restricts forward speed to about 6 km per hour; with heavy bales the work-rate may also be progressively reduced as the operator tires. Thus unmanned sledges and collectors are preferred. There are two main types of these: random collectors, and fully-automatic bale sledges which form the bales into groups which can then be picked up by a tractor loader without manual handling.

Most random collectors consist of metal cradles, which group up to twenty bales, depending on the height of the sides, and allow the tractor driver to form neat windrows (Plate 6.1). They are cheap and uncomplicated, but the bales tend to be handled roughly and, particularly on flinty soils, strings can be broken or pulled as the bales are dragged along. An improvement is the wheeled carrier which collects the bales via a chute either on to a platform or into a V-shaped container. This equipment costs up to three times as much as the simple ground collector, but it does reduce bale distortion and damage, and leaves the bales in a more compact heap for subsequent handling. It is also easier

Plate 6.1   Random bale collector. Sledges of this type collect 15–20 bales at a time, which are dropped as windrows across the field.

to notice if the knotter has failed to operate, so avoiding the problem of untied bales.

Wherever possible the bales are deposited in rows along the field. Very rapid collection, at over 300 bales per man-hour, is possible from such rows by using a large low-platform trailer on to which the single bales from the group are loaded manually; the ideal working height is between 0.6 and 1.6 metres from ground level.

Random-grouped bales can be built into field heaps by hand at up to 450 bales per man-hour. This can be worthwhile in terms both of the weather protection provided by the heaps, and of the unit loads they form for subsequent handling operations. The bales from these heaps can of course be loaded singly, using an elevator, and two men can load into a trailer at about 190 bales per man-hour. But it is much better, once bales have been grouped in this way, to handle them as a complete unit from field to store with a fore-end loader system.

An automatic accumulator eliminates the need to form bales into unit loads by hand (Plate 6.2). The most common form arranges bales into

Plate 6.2 Automatic bale sledge. Models are available to form bales into flat layers of 8 or 10 which are automatically discharged into a uniform heap, ready for handling by a suitably designed loader.

single flat layers of eight or ten bales; several versions are now made, and in some the complexity and cost has been reduced by forming the groups of bales on the ground. If the hay is in a suitable condition these groups can then be moved direct to the barn; alternatively they can be stacked into heaps of 32–100 bales, at a rate of up to 400 bales an hour, using a fore-end loader.

Many attempts have been made to develop fully-automatic accumulators which will build larger heaps of bales suitable both for weather protection and for subsequent tractor fore-loader handling. However, as with the machine just noted, the heavy weight of the combined baler and accumulator could become a disadvantage, particularly on sloping land, and under wet conditions when it could cause sward damage.

Thus if larger units are needed for handling from field to store they are more likely to be built up from smaller groupings of bales than directly. For this purpose a wide range of loaders have been developed to deal with the various groupings of bales. Single layers of bales are generally impaled from above for lifting (Plate 6.3), while stacks of more than one layer are handled by side-gripping devices. The rate at

Plate 6.3   Impaler loaders can handle bales in either flat-8 or flat-10 formation, ideally loading onto a large flat-bed trailer.

which trailers can be loaded and bales hauled to the barn depends very much on trailer size and positioning and by the size and distance between the heaps. Squeeze-type loaders can handle between 250 and 500 bales per hour. Impaler loaders, also operated by one man, can load up to 500 bales an hour, or even more if the trailer is large enough to take four units of eight bales in each layer.

Special-purpose carriers can be used to transport bales in preformed stacks which are picked up and set down rapidly by a single worker and tractor. The tractor-mounted carriers use a buckrake principle with tines which slide under a stack of sixteen bales, the unit being tilted back during transport and the bales held in position by a light frame. Using hydraulically-squeezed sides a load of forty bales can be securely carried. In both cases a further twenty bales can be carried with a foreloader attachment. For safe operation however the tractor must be loaded and driven well within the limits of stability; thus the more rugged rough-terrain loader can be equipped to carry up to sixty-four bales. However, transit distance can limit output, which falls off markedly with tractor-mounted systems carrying less than fifty bales when distance exceeds 1 km. Trailed carriers, using both side-gripping and tilting principles, can travel faster; handling stacks of 50 to 130 bales or more they offer a very efficient transport system.

However, bale carriers can only successfully handle stacks of bales which are upright with straight sides, and which fit within the extended frames of the carrier. Serious problems can be caused by stacks which have been poorly constructed, or which have distorted by settling. This is most likely to occur if the small stacks are handled more than once. Thus systems have been developed in which the block of bales is banded together immediately after it is formed. Packs of twenty bales are banded automatically in the unit trailed behind the baler, and can be handled several times without damage. After banding the packs can be rapidly loaded into trailers and unloaded at the barn, using a hydraulically-operated clamp-type loader. Overall work-rate compares well with other unit-load handling systems, and one man can cart up to 3,000 bales in a normal working day. The packs can also be recovered mechanically from store for feeding. However, due to the weight and complexity of these machines, and their overall length when coupled to the baler, this system has not been widely adopted.

Loading into Store

Mechanisation of bale-handling has concentrated mainly on clearing fields as quickly as possible so as to reduce weather risk. There has been much less mechanisation of handling at the store—generally limited to use of an elevator to move the bales from ground or trailer level on to the stack. Labour use has been high, and work-rates have seldom been matched to the field collection and transport handling systems. In fact up to half the total labour cost has often gone into loading into store. Provision of extra trailers, or the dumping of heaps of bales on the ground by the store, can of course prevent this final stage restricting the rate of removal of bales from the field, but involves yet further expense.

Rate of hand-loading using an elevator depends on how easily the bales can be picked up. If they have been tipped in random heaps they can be transferred to an elevator at about 400 per hour, but this can be increased to over 500 bales per hour if the bales are taken directly from trailers or from evenly set down rectangular heaps. With a further two men needed to stack the bales from the elevator the respective outputs are about 140 and 170 bales per man-hour.

Where the layout is convenient fore-end loaders are now being more widely used for this operation of loading into store. To be fully effective the unit-load dimensions should match the bay-size in the barn, and the loader should be able to stack bales at least fourteen layers high. Using an impaler-loader to handle flat-8s, nearly 500 bales an hour can be unloaded from trailers and stacked into a barn by one man.

Overall System Performance

Many different combinations of equipment may be included in the complete handling system, and provided the output of the successive stages is well matched, or at least that a bottleneck in one stage does not reduce the rate of working in another, bales can be very rapidly cleared from field to barn.

While the work-rate in individual parts of the system can be spectacular the overall output from field to store is often very similar with different systems—although output per man-hour *can* be considerably improved by use of the more expensive equipment. Thus the figures given below indicate average rates of working, accepting that much higher rates can be recorded in some cases.

With a simple low-cost system comprising a bale-collector, 4-wheel trailer and elevator, 100 to 120 bales an hour can be hand pitched and loaded by three men, transported about half a mile and unloaded by elevator to the stack; output is 33 to 40 bales per man-hour. By substituting two special-purpose low-loading trailers for the slightly cheaper 4-wheel trailer, two men can handle up to 125 bales an hour, increasing output to 62 bales per man-hour. A tractor fitted with front and rear bale-carriers can move 85 to 100 bales an hour by direct haulage over the same distance. Two men may be needed for part of the time, but output is likely to be at least 50 to 60 bales per man-hour. Using more expensive equipment, such as an automatic sledge and impaler-loader with a large 4-wheeled trailer, one man can load, transport and stack up to 100 bales an hour.

Large Bales

The tractor foreloader has a lifting and carrying capacity which equips it to handle single large stable units, and this has directed attention to the production of large bales for hay and straw, and more recently for silage.

The first effective models of large balers were in fact designed by farmers, and that developed in 1967 by two Gloucestershire farmers, Pat Murray and David Craig, was the main one in commercial production for the next ten years. This produced rectangular bales, 1.5 m × 1.5 m × 2.4 m, weighing 250–500 kg, and made up of a series of bundles of hay, arranged in parallel across the bale and held together by three heavy-duty polypropylene twines. This 'bundle' formation, together with the low density of 80–100 kg/m³, makes this type of bale very suitable for both natural and forced-air ventilation, and large amounts of hay have been dried in these bales in barn-driers, for which they are much more suitable than the roll bales considered below. They are also more open to penetration by rain than round bales and should be moved under cover as soon as possible after baling. Correct swath preparation is vital if the bales produced are to be regularly shaped and remain stable when handled several times. A 'gripper' attached to a foreloader is used for handling rectangular bales; this has arms which fit along each side at the bottom of the bale and close hydraulically.

Overall baling rates of 6 to 9 tonnes per hour are typical, and these bales can be transported by trailer and loaded into store at a rate of at least 3 tonnes per man-hour. Stable stacks are easily constructed with the bales stacked on their flat sides against a retaining wall. The store must allow access by a foreloader; feeding to stock then presents few problems, and the hay in each bale tends to remain in the original bundles which are easily distributed to stock.

However, the more recently-developed round bales are now the most commonly seen. The many models of large round baler are of two main types:

Fixed chamber (Plate 6.4) balers in which the diameter of the bale-forming mechanism is fixed. In the initial stages hay is randomly packed into the chamber. Rolling action starts when the hay comes into contact with the bale-forming mechanism, which can be based on

Plate 6.4 A 'fixed chamber' roll baler uses a system of ribbed rollers to form bales of fixed diameter (Photograph Power Farming).

rollers, chain and slats, or belts. This completes the bale with a spiral of compressed roughly aligned crop. The outer layers of bales made by this type of machine are thus more dense than the core; bale size is fixed by the diameter of the chamber.

Variable chamber balers (Plate 6.5) produce bales in which the density is more regular across the diameter of the bale. The bale-forming mechanism is a system of belts or chain and slats which expand with the bale, which is thus subjected to continuous rolling. Bale size can be determined by tying at virtually any stage of the process.

These machines are available with bale chambers varying in size from 0.9 m diameter $\times$ 1.2 m wide to 1.8 m diameter $\times$ 1.5 m. Bale density is likely to be about 110–120 kg/m^3 with dry hay, the bales weighing from 0.2 to 0.5 tonnes depending on size. These balers are relatively simple in operation and construction, but the shape and stability of the bales made depend very much on the shape of the swath picked up and the skill of the operator. Bales are simply wrapped with thin twine and then ejected—an operation that can occupy 30–40 per cent of the baling cycle. Work-rate ranges from 6 to 10 tonnes per hour, depending in particular on crop yield and bale size.

Plate 6.5   A 'variable chamber' roll baler uses a system of belts to form the bale, which can be varied in diameter (Photograph Power Farming*).*

These large, dense packages are not really ideal for hay as heat and moisture are not readily dissipated from them. Thus the hay should be drier than is considered safe for standard bales, and preferably close to 18 per cent moisture content. In theory bales with the lower density core should be at less risk to heating. However, artificial ventilation of round bales on any scale is much less effective than with rectangular bales, as the airflow must be directed along the axis of the bales. It has also proved difficult to distribute preservative chemicals (page 13) uniformly in round bales, and experiments at Silsoe indicate that two to three times as much ammonium propionate is required to preserve hay at 25 per cent MC in large round bales as for hay at 30 per cent MC in standard bales.

Because of these problems in safe storage, large round bales are not now widely used for hay, although some farmers, particularly in the north, have used round balers, purchased for making silage, to make hay from relatively light regrowth crops cut in mid-summer. The bales are left exposed in the field to dry by ventilation before they are stored, taking advantage of their resistance to weather which allows them to withstand some rainfall without damage. However, by far the commonest use of round bales is for storage of straw, and these bales are convenient for subsequent treatment with ammonia (page 146). Some models can apply a loosely woven polypropylene net in place of the twine, thus speeding up this part of the operation, forming neater bales and adding to their weather protection.

Roll bales are handled at all stages by simple tractor foreloader attachments and transported either by trailer or bale-carriers, using principles similar to those for standard bales. There may be difficulty in making full use of the storage space in existing barns because of problems of access for the loader and of constructing tall and stable stacks of round bales. Thus these bales are most easily stacked in pyramid fashion, but this does waste much space.

BARN HAY-DRYING

We have noted the considerable advantage there should be in completing the drying of hay in the barn rather than relying on natural drying in the field, with the attendant risks of loss of yield and of nutritive value. Much of the development work on this technique during the last thirty years has been carried out by the electricity supply industry, which has published detailed advice on installations and techniques. Advice has also been available from manufacturers of specialist drying equipment as well as from the mechanisation advisers of the advisory services. Yet despite this support, and the widely acknowledged benefits, barn-drying techniques have been slow to gain farmer acceptance, and it is probable that less than 4 per cent of all the hay in the United Kingdom is made in this way.

There are several possible reasons. Barn hay-drying systems which significantly reduce field exposure time generally require considerable capital investment; the systems always involve some manual handling and stacking of bales; power and fuel costs have increased. Probably of more importance, alternative systems of conservation, most notably reliable silage techniques with low capital requirement, are now being adopted on farms on which hay was previously made. Nevertheless the ability to remove a partly dried crop from the field and complete drying in the barn so as to avoid either excessive leaf shatter or damage from rain is worthy of consideration by the many farmers who continue to make hay.

Systems of Drying

Artificial drying can be divided into two broad categories. The first of these, *conditioning*, is simply a method of ventilation in which enough air is passed through the crop to prevent heating, and at the same time gradually removing a small amount of residual moisture—less than 300 kg per tonne of finished hay—from the crop over a period of ten to fifteen days. This keeps respiration losses to a minimum; it also helps to prevent the *natural* convection, aided by heating, which removes moisture from the crop in one part of the store and deposits it in another, generally in the top layers, where it can set up moulding. Thus conditioning must be regarded as a process, *attached to the end of an existing method of hay production*, which reduces the risk of excessive loss in the final stages of drying; it is best suited to dealing with crops of 35 per cent moisture content or less. Drying continues as long as the relative humidity of the ventilating air is less than the equilibrium value equivalent to the moisture content of the hay (Table 6.1). Hence unheated air of less than 90 per cent humidity will not only keep hay cool, but will also dry it to below 30 per cent moisture content. The usual recommendation is to ventilate the hay continuously for the first few days in store to keep it cool, and from then on to switch the fan off at night and only to operate it when air conditions are favourable for drying.

Table 6.1   **Equilibrium values for hay moisture content and air humidity**

Relative humidity of air (%)	Hay moisture content (%)
95	35
90	30
80	21.5
77	20
70	16
60	12.5

The second broad category is *drying*, in which sufficient air, as well as added heat, is used to remove between 300 and 700 kg of water for every tonne of dry hay (15 per cent moisture content) that is produced. Depending on the system chosen, moisture content at loading from the field will be between 35 and 45 per cent, levels at which hay will not store satisfactorily, without ventilation, even if it has been weathered in the field for a considerable time. Drying is thus *an integral part of the haymaking system*; without it a storage loss of at least 15 per cent could be expected.

Drying must take place sufficiently quickly so that the *drying zone* (the area in which water is being transferred from the hay to the air) passes completely through the hay before the top layers start to deteriorate. If the air is very wet it must be heated to increase its water-holding capacity. Detailed Tables are available for a range of conditions; but, for example, if the atmospheric air is at 16°C with a relative humidity of 90 per cent, raising its temperature by 2.7°C will increase its drying potential five-fold. However, because of non-uniform airflow and resistance to moisture loss from the crop, especially from bales, the actual gains may not be as impressive as this. As a rough guide a temperature rise of 5.5°C under adverse ambient conditions will improve drying rate to the equivalent of that expected on a fine summer's day.

There must also be a careful balance between the temperature of the atmosphere and that of the exhaust air from the drier; if the air passing out of the crop has too high a temperature it will not be saturated, energy use will be inefficient, while the hay at the bottom of the drier may be seriously overdried; but if the air emerging is completely saturated and cooled it will deposit water in the top layers of hay, which may then mould. In practice it will generally be possible to maintain the humidity of the exhaust air at 90 per cent during the early stages of drying. But, as drying proceeds the humidity of the drying air should be progressively reduced for, as can be seen from Table 6.1, air at 90 per cent RH will dry hay to 30 per cent MC, but air at 65 per cent RH is required to dry hay to 15 per cent MC, the safe storage level. This can be done by selecting good drying conditions if the hay is not deteriorating during the later stages of drying or by adding only sufficient heat to correct the humidity to the required level.

Control of airflow at the recommended levels and, where heat is used, restriction of heat input so that the temperature of the exhaust air is not more than 3°C above ambient, is likely to give the most satisfactory and economical drying.

Choice of Drying System

The biggest single factor influencing the choice of system for a particular farm must be the amount of wilting that experience suggests can reliably be done in the field, and therefore the amount of water that remains to be removed by artificial ventilation. Drying from a high moisture content, above 55 per cent for example, although technically possible with the fan and heater equipment commercially available for barn-drying, would be inefficient and costly.

In order to fully exploit the potential of the system, hay for barn-drying will generally be mown at a less mature stage of growth than for field-hay—probably when most of the ears have just emerged from the stems, yet when field-wilting to below 50 per cent moisture content is not too difficult. Barn hay from even less mature crops, cut when few ears have emerged, *can* be made but presents greater problems; an extended period of field wilting may be needed, and ventilation must be continued for some time to make sure that the hay does not subsequently heat up through continued oxidation of its sugar content, leading to moisture condensation in the upper layers.

Whichever system of drying is chosen, and particularly with bales, which have a much higher bulk density than loose or chopped hay, it is recommended that the mean moisture content of the stored hay should be reduced to 15 per cent, to ensure that there are no wet patches left to act as a nucleus for heating and moulding.

Storage Driers
For most farms some kind of storage drier is likely to be the most suitable. This has a higher capital cost than the small, or the very simple, batch drier, but a lower running cost. The overriding consideration is that the bales do not have to be moved again after they are dried until they are fed. Because batches of hay can be loaded progressively, each batch being dried over a period of two to three weeks before the next batch is loaded, these driers can operate mainly with unheated or only slightly heated air, which prevents over-drying; the moisture content of hay collected from the field should be in the range 35–45 per cent, depending on the exact specification of the drier. This means that hay will generally be wilted for two to five days in the field, and can be baled from the swath some one to two days before it would be considered fit to bale for normal storage.

Most existing barns can be quite simply modified for use as storage driers, but some aspects of design are critical, so that it is prudent to obtain advice or to consult one of the specialist booklets on the subject. Storage driers can be fitted into any building which is suitable for storing hay, but the building should preferably have a height to the eaves of not less than 6.6 metres; a Dutch barn is particularly suitable (Plate 6.6). Reasonably airtight side-walls are required to prevent air escaping from the sides of the stack. These should be constructed to within 0.5 metres of the maximum height of loading, and should be fitted with access doors 1.8 metres wide for loading and unloading. The dimensions of the drier will often be dictated by existing building dimensions. A bay 9 m × 4.5 m with a 5.5 m working height will hold between twenty-three and thirty-two tonnes of dried hay.

The ventilated floor should be constructed to take a load of 1,200 kg per m^2. Welded mesh is often used, supported 0.6 metres above ground level and with an area 0.6 metres wide around the perimeter blanked off to prevent air escaping between the bales and the walls. Fans are

Plate 6.6 Most buildings suitable for storing hay can be converted to 'storage' drying by constructing a carefully-dimensioned drying floor and side walls.

normally sized to provide an airflow of up to 0.25 metres per second per square metre of floor area ventilated, at a static pressure of 600–750 N/m^2 (60–75 mm water gauge).

For this type of storage conditioning hay is best wilted to between 35 and 45 per cent moisture content before it is loaded. Ideally the bales should be packed tightly together in the drier, without gaps which would allow direct passage of air. Four or five layers of bales may be loaded on the first day, the fan being switched on as soon as the first layer has been completed. After ventilating for between one and four days the drier can then be topped up with two or three layers at a time at intervals over a period of three or more weeks. As many as sixteen layers can be loaded finally, but the moisture content of bales loaded above the twelfth layer should be reduced below 35 per cent before loading, that is to a level at which hay is often baled and stacked in small heaps in the field. After loading has been completed up to ten days further ventilation may be necessary, but during the last week the fan can be switched off at night.

In practice it is quite difficult to determine when drying has been completed. The safest technique is to leave the fan off for a couple of days once it is thought that the hay is dry, and then to run it, immediately checking for any smell of warm, damp hay in the exhaust air. This process is repeated at gradually increasing intervals until there is no doubt that the hay is safe to store.

Unwalled Storage Driers

Many attempts have been made to reduce the capital cost of storage-drying installations, particularly by omitting the side-cladding and the perforated floors. In these driers the air is distributed more crudely by a system of ducts across the floor of the barn; these are built of simple materials such as breeze-blocks and timber, and should have a cross-sectional area of 1.0 m² per 10 m³ per second of air delivered by the fan. The spacing of the ducts is determined by the final height of the stack to be ventilated, and the distance from the ducts to the outside of the stack should be the same in all directions in order to even out the airflow. But in practice it is impossible to avoid uneven air distribution, and this type of drier cannot be expected to condition hay loaded at much above 30 per cent moisture content. Even then success depends on adequate fan capacity and carefully designed ductwork.

Drying Large Rectangular Bales

As already noted, techniques have been developed for drying large rectangular bales (page 94). These are formed into arches, each of eleven bales, with a central duct aimed at distributing air 'radially' through the stack (Plate 6.7). Carefully dimensioned bale supports, positioned in the duct space as the stack is being built, keep the ducts open as the stack settles during drying. Bales can be dried down from 30–35 per cent moisture content in ten to fourteen days, using an airflow of 0.35 m³ per second per bale at a static pressure of 400 N/m² (40 mm water gauge). These tunnels do not utilise barn space efficiently so that the bales, once dry enough for long-term storage, are often restacked. This emphasises the importance of making firm bales with a regular shape which can be formed into a tight and stable tunnel, and then stand up to further handling when dry.

Handling and Drying Loose Crops

Loose-hay drying systems have not been much used in the United Kingdom, although they have been widely used in parts of Europe, and particularly in Scandinavia. In those countries hay is traditionally stored close to, and often above, the stock, so that feeding loose hay presents fewer problems than it would with the larger herds and more extensive building layouts typical of the UK.

Systems are now available, however, in which loose hay can be

Plate 6.7   A drying tunnel of large rectangular bales; each 'arch' contains 11 bales.

mechanically handled at every stage, including extraction from the storage barn and transport to the feeding point. Crop is picked up from the field by a self-loading trailer, coarsely chopped, and fed into a pneumatic conveying system via a dump-box. Even distribution in the barn, but without compaction, is critical. This is achieved by a telescopic pneumatic distributor suspended from the roof apex of the barn (Plate 6.8). Material at up to 50 per cent moisture content is loaded in shallow layers over a ventilated floor where it is dried *and* stored; after drying leafy hay settles to a density of 110–130 kg/m³. The hay is unloaded from the store either by a foreloader and grab, or by a recently developed block-cutter, which extracts 2 m³ blocks of hay, weighing 200 kg or so, for transport to the feeding point. Because 'drive-on' floors are expensive to construct, the ventilated floor can be made up of sections which are removed as the hay is taken out from a near-vertical face by the unloading vehicle working at ground level.

These techniques are well-developed on many European farms, where installations have been designed with airflow and temperature rise matched to the high loading rates and moisture contents. With skilful operation very high standards of conservation can be achieved— well above those generally associated with hay—but these must be evaluated against the fairly high capital costs and management skill needed.

Plate 6.8   Telescopic pneumatic distributor used to fill a barn with loose hay, evenly and without compaction.

Many farmers will continue to make hay, in spite of the difficulties, because they consider that it is a good and predictable type of feed for their stock and, especially if barn-drying is used, that it will give them good production from their grassland. But the main limitation of most haymaking methods at present in common use is that, because of their extreme weather dependence, there is no guarantee that the crop can be harvested at its optimum stage of growth, particularly if the rate of mowing must be geared to the throughput of a barn hay-drier.

One of the most important reasons for the increase in silage-making, at the expense of hay, must be that the handling and storage methods employed for silage allow the bulk of the grass crop on most farms to be cut and stored within a period of ten to fourteen days. Particularly where only a small amount of wilting is required, as in most current silage systems, the conservation programme can then be planned and carried out to schedule in at least four years out of five.

Chapter 7

SILAGE MAKING

INEVITABLY THE DAY will come when the silage-maker faces the fact that his consideration of techniques is at an end, and he must get down to doing the job. He should have weighed up such factors as the type of crop he will grow, the animals he wishes to feed, and the farming layout within which he must work, so that certain practical priorities will be clear in his mind.

These will be influenced by the theoretical considerations we have outlined, but it is undeniable that many a farmer finds it difficult to plan and organise a silage-making programme in the light of the conflicting exhortations of his various advisers.

The following six key priorities will assist in clearing the mind and arriving at workable conclusions:

1. Cut the crop at the optimum stage of growth.
2. Achieve, if possible, a minimum crop dry-matter content of 22 per cent, and apply an additive as necessary.
3. Ensure that the crop coming into store is chopped to the correct length.
4. Eliminate the chance of soil contamination at any stage.
5. Fill the silo by a method that prevents air movement, heating and oxidation.
6. Completely seal the silo against entry of air as soon as filling has been completed.

HARVESTING THE CROP

Stage of Growth at Cutting

Enough has already been said to underline the importance of this factor, and the establishment of the characteristic D-value curves for most varieties of crop has been appreciated by many grassland enthusiasts;

yet failure to comply in practice with this recommendation is still wide-spread. The common temptation to wait a few days for a 'little more bulk' will seriously affect crop quality. The fact that the stage of growth of the crop at cutting will have more influence on the eventual feeding value of the product than any of the other factors under the farmer's control accounts for the high priority we give to it.

Dry-Matter Content

Most forage crops contain less than 20 per cent of DM at the time they are cut, and there are several advantages in raising the DM content of such crops before they are loaded into the silo, including the reduced weight of crop that has to be carted from the field, the probability of a more stable fermentation, a reduced quantity of effluent flowing from the silo, and in some cases improved nutritive value in the resulting silage.

During the 1970s the reported success of the high-dry-matter silage system being used in the Netherlands encouraged a trend towards more extensive wilting in the United Kingdom, with dry-matter levels above 35 per cent being advocated. However, it soon became evident that this level of wilting was practicable in Holland because most of the silos held less than 50 tonnes of forage, and could be filled in only one or two days, often by large contractors' machines: thus it was possible to wait for a dry spell before cutting. The problem of wilting is quite different in the United Kingdom, where many farms make upwards of 500 tonnes of silage in ten days, so greatly reducing the probability of good weather for wilting. To insist on extensive wilting *must* reduce the chances of the crop being cut at its optimum D-value, in particular in the higher-rainfall grass-growing areas of the west and north. Local weather records provide information on the monthly distribution of rainfall, but of more importance are the likely hours of sunshine, since it is the incidence of drying conditions which matters rather than the absence of rain. Much of the grass cut for conservation in Britain will be harvested during May and early June, and weather records during this period make clear how difficult wilting can be. It is also the time when the moisture content of crops is at its highest. Although it is generally possible to increase the dry-matter content of a sappy crop on the day of cutting by applying a severe conditioning treatment, two consecutive fine days are often needed to increase DM to the range 25–30 per cent, and then only when mechanical treatment is given immediately after the swath is mown.

Thus the most general advice now is that there is advantage in a limited degree of wilting for up to twenty-four hours, aiming at a dry-matter content of 25 per cent in the crop going into the silo; but if the weather has not allowed this degree of wilting the crop should then be ensiled, if necessary using an additive, rather than leaving it on the field to suffer further losses, at the same time disrupting the operation of the harvesting team. The main reason for selecting silage-making in clamps or bunkers as the preferred method of crop conservation is that these provide the most effective way of ensiling crops in the 20–30 per cent dry-matter range, and so are the most independent of the need for a spell of fine weather. This is embodied in the 'minimum wilt' concept, in which cutting takes place as near as possible to the date indicated by the D-value curve, while a target of DM towards the lower end of the scale, in the 22–25 per cent range, is accepted. At this level of wilting there is a useful reduction in the weight of the crop that has to be handled, little effluent will flow from the silo, and the time that the crop is at risk to the weather in the swath can be reduced to twenty-four hours or less.

There *will* be occasions when favourable weather coincides with harvesting, and the level of dry matter we are suggesting will easily be achieved, but under these conditions it is then necessary to keep tight control to avoid over-drying in the swath. By adjusting the interval between using the cutting machine and the harvester a good measure of control is possible, though varying weather conditions and overnight dews can make this quite a difficult exercise.

Chop Length

Generally speaking the shorter that forage is chopped during the harvesting operation the more of it can be packed into the trailer and the better it consolidates during the crucial period when the silo is being filled. Short, regularly chopped silage creates fewer problems with unloading and feeding machinery, and is essential for loading and unloading tower silos: it also has nutritional advantages, as animals can eat more of short-chopped than of 'long' silage, particularly when self-feeding from a silage face.

The degree of chopping required will depend upon several factors, but it is generally agreed that material 25–50 mm in length suits most requirements. Cut forage above 30 per cent DM is more difficult to consolidate and failure to expel air trapped within the crop can lead to heating and mould formation. Consolidation is aided by chopping and the drier the crop the more beneficial are the effects of chopping and the more critical is the need for effective sealing. Table 7.1 indicates the maximum chop length appropriate to crops at different dry-matter contents.

This illustrates the point that the drier the crop the shorter is the chop length needed to ensure that air is expressed from the crop in the silo;

Table 7.1   **Maximum chop length appropriate at different dry-matter contents**

Target DM (%)	Maximum chop length (mm)
under 20	200
20–25	130
25–30	80
above 30	25

the better consolidation possible with short-chopped material makes heating and poor fermentation much less likely.

Traditionally the chop length of forage has been described in terms of the machine used to harvest it. With the proliferation of designs and the range of chop lengths at which most machines can now be set this type of classification is no longer adequate. One method of expression is in terms of the *median chop length*. For example, in *a crop sample with a median chop length* of 25 mm half of the sample by weight will be more than 25 mm and half by weight less than 25 mm in length. This provides the basis for a general classification using a single figure. In some circumstances, however, it is more important to know about the quantity of material at certain chop lengths within the sample; for example, when considering the handling and flow properties the proportion of material above a certain length (say 100 mm) may be critical. It is important therefore to describe chop length in relation to the particular crop and silage system.

Types of Harvester

The relationship between the storage system, the dry-matter content and the length of chop required will largely determine the type of harvester to be used, and how it is to fit into the silage system selected.

In the early stages of the mechanisation of silage-making the flail forage harvester was preferred for its simplicity, low capital cost and flexibility. In recent years the inherent disadvantages of this design have become more apparent as alternative designs have improved.

One of the drawbacks of the simple flail harvester is the tendency to exert a powerful suction effect on the crop and on adjacent ground areas, which can lift soil with the forage. This may lead to serious soil contamination which can upset subsequent fermentation. This undesirable effect can be aggravated if the machine is set to cut close to the ground so that the spade-shaped flails scalp the sward and pick up earth and small stones. The main disadvantage, however, is that few of these

Plate 7.1   The double-chop harvester either cuts and lifts the crop, or picks up crop from the swath, with cranked free-swinging flails. The crop is transferred by a cross-auger to a flywheel-type cutting head.

machines can be adjusted to chop shorter than a median length of 150 mm, so that a fair proportion will exceed 250–300 mm. This longer material limits the amount that can be loaded into a trailer, interferes with compaction, and makes control of air movement in the silo more difficult. Silage made from this long material is also difficult to extract from the silo, and may restrict intake of animals when self-feeding.

The double-chop harvester (Plate 7.1) is more complicated than the simple flail. It can perform the same three functions of mowing, mowing and loading, or it can pick up a wilted crop, although the chop length produced by the edge-cutting curved flails when it is used as a mower may be too short to allow efficient pick-up and loading after wilting. So, if this harvester is to be used for picking up from the swath, the best results will be obtained by mowing with another machine, designed for the job, which will leave forage of 200–250 mm length set up in a swath of satisfactory shape for wilting and loading.

This type of harvester, which can cut and harvest direct as well as pick up cut crop from a swath, is most useful; a convenient and effective working procedure is to load the crop mown on the previous day until the middle of the afternoon, by which time the standing crop is surface-dry and can be cut direct; this will reduce the work-load on the pre-cutting machine and can avoid the harvesting of over-wilted crop. It also permits harvesting to continue uninterrupted if the pre-cutting machine breaks down.

The double-chop harvester produces material in the range 50–150 mm, which greatly increases the weight of crop per trailer-load; the resulting forage is well-suited for efficient filling and settlement in surface silos, for self-feeding, and for mechanical unloading from the silo. The design and arrangement of the cutting flails result in minimum suction being applied to the crop, and if the cutting height is correctly adjusted there is less risk of soil contamination than with a machine with spade-shaped flails.

The various points in favour of shorter-chopped material have led to increased use of 'precision-chop' type harvesters with surface silos (Plate 7.2). The higher initial cost is compensated for by a big increase in potential output, which can be of crucial importance to the large-scale operator—although this higher output will not be achieved unless the machine is matched by commensurately high capacity in both the pre-cutting and hauling elements of the harvesting team. The drive-shaft and gearbox of the precision-chop harvester are capable of transmitting considerable power to the cutting mechanism, and this feature—resulting in high output—demands a powerful operating tractor.

When selecting a tractor the following factors should be considered in addition to the power required by the harvester alone: the power required to propel the tractor, harvester and possibly loaded trailer; the effect of slopes and ground conditions, and the ability to select a forward speed which matches the size of the swath to the harvester output. Halving forward speed halves the power required for propulsion, while climbing a 1 in 10 slope requires three times as much power as maintaining the same speed on level ground. Towing a loaded trailer in addition to the harvester can require 15 per cent more tractor power than towing the harvester alone. In order to maximise harvester output a tractor is needed with sufficient power to meet both these requirements

Plate 7.2   The cylinder-type 'precision chop' harvester picks up the cut crop and delivers it, via feed rolls, to the chopping cylinder.

as well as to power the harvester. This is most likely to be achieved by picking up a large, even windrow at a slow forward speed. Selecting the forward speed which best matches crop pick-up rate to potential harvester throughput can be critical and requires a close range of gear ratios at low speeds. Where the harvester is equipped with its own diesel engine the forward speed of the tractor can be continuously adjusted to match changes in ground conditions and crop without affecting the power available to the harvester. A more consistent output should therefore be achieved. Aspects of power utilisation and chop length relating to various types of harvester have been reported by ADAS in 'Silage Harvesting; Power Utilisation', 1982.

The shorter chop, median length 35 mm, typically produced from the precison-chop harvester, allows much more dry-matter to be carried on each trailer, an important aspect where long hauls are involved, as is often the case in harvesting operations requiring these higher-output machines. The chopped material is also easy to handle with loading machinery at the silo, and consolidation of the cut crop is very rapid, giving excellent control of air movement and heating. The silage can readily be self-fed, or loaded by front-end loaders, and will flow more uniformly than long material in mechanised feeding systems.

Because precision-chop machines are used mainly for lifting wilted herbage, they are normally equipped with a pick-up attachment. On stony land the pick-up tines tend to lift stones with the crop, and these can cause damage to the precision-chopping cylinder. This is most liable to happen where stones have been moved into the swath in the process of collecting two swaths into one, and serious delays can result. So common is this problem on some soils that harvester operation has to be limited to the single width of the pre-cutting machine, with consequent reduction of output potential: this operation has been speeded up by the introduction of mowers with cutting widths of 2.4 m and over which will set up a large weight of crop in a single windrow.

The now common use of precision-chop harvesters has encouraged a number of variations in design, each with beneficial features. The general objectives have been to improve reliability by more robust construction and in some cases by simplified design. Machines using the more common 'cylinder' type of chopping mechanism have been developed further to minimise the risk of damage—a feature of paramount importance with the concentrated periods of high performance now required. This has been achieved by incorporating metal-detecting facilities, while in some designs the chopping cylinder is made up of a series of individually-mounted blades which 'fail-safe', with less damage overall (Plate 7.3). One manufacturer has reversed the direction of rotation of the cylinder, claiming significant economy in power consumption as a result.

Precision-chop machines are of course essential where the crop is to be loaded into a tower silo, since the unloading machinery will not deal efficiently with the longer, less precisely chopped material from other types of harvester. There are also particular crops, such as maize, which can only be satisfactorily handled by this type of machine.

However, while there remains much advantage in short-chopped forage in both the making and the feeding of silage in bunker silos, the trend is towards less precise specification of the chop length. Thus there appears scope for simplifying the various chopping mechanisms by eliminating the feed rolls previously required to achieve precise chopping. The effect of this change in design is most apparent at very short chop length (5 mm); the median chop length without feed rolls is then likely to be three times the set chop length, compared with only twice from a machine with feed rolls. However, when the chop length is set to the more usual 25 mm there is less difference between the samples produced, each having a median approaching the set chop length. The need for less precise chop length, coupled with improved knife maintenance, has also encouraged the reintroduction of the flywheel chopping mechanism, well known for its ability to absorb the fluctuations in

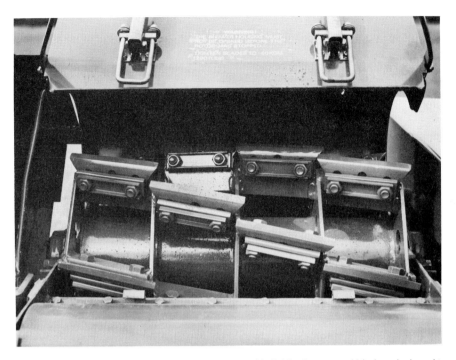

Plate 7.3   A chopping cylinder fitted with a series of individually mounted blades, designed to 'fail-safe' with minimum damage to the machine.

power demand arising from the uneven pick-up from the typical swath (Plate 7.4).

Forage Harvesting Systems

In practice the output of a forage harvesting system is frequently limited by the transport facilities, and the importance of this part of the system cannot be over-emphasised. Much can be attributed to detailed aspects of preparation and planning, which, though obvious, are easily over-looked. Although chop length has a bearing on trailer capacity other 'local' factors have a major influence, such as the skill of the operator, crop dry-matter content, and the distance travelled to collect a load.

Single-cut flail harvesters are of two types: those designed to fill a trailer towed integrally behind the harvester, and those which load into a separately-towed trailer running alongside. The former has the advantage that a second tractor and operator are not required; further, although there is often a higher output per hour from side-loading,

Plate 7.4   Harvester with a fly-wheel chopping mechanism which absorbs fluctuations in power demand.

because the job of coupling and uncoupling the trailer is avoided, the output of in-line systems can be increased by using a hydraulic pick-up hitch. However, on sloping fields and soft land there is considerable advantage in the separately-towed trailer, since the tractor-harvester-trailer train can get into trouble through wheel slip, bogging down or overturning.

Some flail harvesters are mounted on the side of the tractor and elevate the crop into a trailer drawn directly behind it. They are particularly suited to operation by one man, as the harvester can be quickly uncoupled, leaving the tractor and trailer free to haul crop to the silo.

Most double-chop and precision-chop harvesters will fill either a trailer towed behind them or one running alongside. In either case when high dry-matter crops are being loaded care must be taken to screen the top of the trailer so that the light and leafy particles are retained.

To prevent losses from trailers, especially when cornering, it is necessary to turn the harvester chute sideways, and to obtain a full even load the flap must be moved up and down to fill the trailer from back to front. The fitting of tractor safety cabs has made the control of trailed machinery and the operation of simple mechanical levers more difficult, a problem that has been overcome by use of electric motors and hydraulics to swivel and to change the angle of the flap above the discharge chute.

Another major cause of delay in the transport system arises when badly designed trailer doors and fittings lead to jamming, so that brute force must be used to open doors to remove the crop. Even when the doors are open, the compacted crop can often only be emptied out by jerking the trailer forward on the tractor clutch. The most effective type of rear door is the up-and-over counterbalanced gate, secured by a spring latch at the base; certainly the ease of opening under load should be one of the main considerations when a new trailer is purchased. For easy emptying there is great advantage in having tapered trailer sides, wider apart at the rear than at the front; standard trailers can be modified quite easily by fitting a smooth lining tapered outwards towards the rear, within the fixed sides. When the trailer is tipped the load slips out easily from between these flared walls and the time saved more than compensates for the slight reduction in the amount of crop carried.

Self-Loading Forage Wagons

Conventional forage harvesters generally have a high power requirement, and because they are difficult to operate on steep slopes when towing a trailer are most efficiently used as part of a harvesting team requiring three or four men. These factors have tended to limit the extent of silage-making on smaller farms in the north and west of the United Kingdom, except where an effective contracting service is available. Self-loading forage wagons (Plate 7.5), originally developed for haymaking and zero grazing, particularly in hilly areas of southern Germany, have proved useful on smaller farms in the United Kingdom. These machines have been progressively improved and strengthened, overcoming some of the mechanical difficulties originally experienced. While some early users, who were making silage for the first time, found that the quality of silage made was poor, advice from both official and trade sources has now led to the development of a system of operation which is well suited to the smaller livestock farmer.

It has been shown that quality can be quite satisfactory, provided that the grass is wilted and a reasonably short chop obtained, and that particular care is paid to even filling and exclusion of air from the silo. These objectives can be achieved by using a machine fitted with twenty or more knives to slice the crop which has been cut with either a mower

Plate 7.5 *Self-loading forage wagons have been progressively improved and strengthened. Fitted with twenty or more knives, these trailers are well suited for harvesting crops for silage when little labour is available and transport distance is short.*

or a mower-conditioner. Although chopping the crop with this number of knives requires little power, typically 6–12 kW, draught requirements may be high in hilly areas—up to 40 kW, compared with less then 10 kW on flat land. Therefore it cannot be assumed that this system only requires a small tractor.

Even so, under the right conditions, the self-loading forage wagon offers the basis of a complete silage-making system with a high output in relation to labour requirements. Because the equipment has a low centre of gravity and less tendency to jackknife than normal harvesters and trailers, it allows the hill farmer to increase the area he can cut for silage by working on steeper slopes, provided that a powerful tractor, preferably with four-wheel drive, is used. The degree of chopping is largely determined by the spacing of the knives, which for design reasons is limited to 30–50 mm. Only a small proportion of the crop is likely to be presented at right-angles to the knives so that up to 70 per cent may exceed 50 mm in length with a median two to three times the knife spacing. This, combined with the absence of laceration, means that attention must be paid to the principles of silage-making outlined in

Chapter 2. Wilting to at least 23 per cent dry matter is important, especially as additives cannot be distributed very evenly. Initial consolidation within the wagon may be considerable and, provided the silo is then loaded evenly and is well compacted, followed by good sealing, fermentation will be satisfactory.

Although this is sometimes regarded as a low capital cost one-man silage system, a second man is needed at the clamp to ensure a quick turn-round. Work-rates of four loads per hour can be achieved when the fields being cut are less than 500 m from the silo—equivalent to 10–18 tonnes per hour, depending on the size of the wagon. At distances greater than 700 m performance falls off rapidly as the proportion of time spent on transport increases.

On some farms two wagons will be needed to keep pace with one man using a push-off buckrake at the silo and, even more important, to complete harvesting whilst the crop quality is still high. Even for much smaller tonnages, a second man may be needed to assist with mowing and at the clamp, to complete the job before quality deteriorates.

These machines have been widely used in Holland, where they have been very successful in harvesting wilted material of higher dry-matter content and longer chop length than is usual for silage-making in the United Kingdom. As noted on page 106, this is at least partly because the silage is typically made in a series of small, well-sealed silos, each of which is both filled and consumed rapidly; this practice limits the deterioration likely if either of these operations is extended.

Harvesting Maize for Silage

Maize quality, unlike that of the grasses, changes little for several weeks after the optimum level is reached (Figure 4.6). This stage, indicated by the grain becoming cheesy and firm and a 'dent' appearing on the exposed surface of the grain, may be reached in southern England any time after mid-September; this date will depend on the variety and the season, and will also be later farther north. At this stage both yield and D-value are high; limited frost will assist in drying out the crop, but will not reduce its feed value.

To ensure satisfactory fermentation in the silo maize *must* be finely chopped, for maize silage readily spoils by heating if large pieces allow air to get into the mass. Thus use of a precision-chop harvester is of great advantage, for although some larger particles will get through, the bulk of the crop will be well chopped. Some machines can also be fitted with a screen to ensure that all the grains are broken. Double-chop and flail harvesters generally give less satisfactory results. Three main types of maize harvester are available:

Plate 7.6   Two-row forage maize harvester mounted on a
'reversible-drive' tractor.

- Purpose-built single- or two-row machines mounted on a 45–52 kW tractor. Output is 10–12 tonnes per hour, or up to 2.4 hectares per working day, depending on tractor power and soil conditions.

- Precision-chop forage harvesters with single- or twin-row maize attachment driven by 52–75 kW tractor, giving an output of 12–22 tonnes per hour or 2.4–4 hectares per day (Plate 7.6).

- Trailed specialist harvesters, either pto or engine-powered, and self-propelled harvesters in the 67–150 kW range. Outputs are 13–30 tonnes per hour, i.e. up to 5.7 hectares per day, while three-row machines have even higher rated outputs. They are well suited to contractors or co-operative groups with good haulage facilities, and their use for harvesting maize in the autumn spreads the capital cost over a longer season.

The fine-chopped material gives heavy trailer loads, a standard 3 m × 1.8 m trailer holding over three tonnes; thus on wet or heavy land double axles or twin wheels are essential, since some of the larger trailers now used can contain up to 8 tonnes of crop.

Avoiding Soil Contamination

The level of soil contamination in cut forage will depend on the type and design of harvester used, and may be increased by mud-splash during heavy rain. Under extreme conditions silage can contain over 20 per cent of soil; but there should be very little soil in grass cut by a correctly adjusted harvester.

This emphasises the importance of correct sharpening and setting of cutting equipment, and particularly of the height of the flails above the ground surface. They must not be allowed to clip the surface, either when cutting direct or when lifting from the swath; as noted earlier, setting the height will be easier if the field was heavily rolled in the spring. Soil contamination can also occur from the tyres of tractors buckraking crop into the silo; the provision of a concrete loading area with a 1.2 m retaining wall at one side against which the buckrake can push is a well worthwhile investment, and with outdoor silos will also markedly improve either self-feeding or tractor unloading of the silage in the winter.

FILLING THE SILO: THE DORSET WEDGE METHOD

Walled Silos

The subjects of stage of growth at cutting, dry-matter content and crop specification have, so far, been considered in rather abstract form. But there is often cause for much heart-searching on the day the decision is taken to start harvesting. Simple matters left undone until that day, such as the overhaul of harvesters, the preparation of trailers, and a check on the unloading gear, can be the cause of quite unnecessary delay. It is remarkable how often farmers leave silo preparation and the attempt to purchase polythene sheeting until the very morning cutting is due to start. For with all the equipment and the silo fully prepared during the previous weeks the whole sequence of cutting and harvesting can swing into action on the date intended.

Plate 7.7   Method of fixing a 2 m strip of 300 gauge polythene sheeting to the silo side-wall, to ensure adequate sealing of the shoulders of the clamp.

The polythene sheets used to control wastage at the wall surfaces should be attached to the silo before filling commences, except where the walls are already airtight, e.g. with rendered concrete or resin-bonded plywood facing. In sleeper-walled silos the side-sheeting should come down to ground level to prevent any air entering.

Side-sheets to control shoulder wastage are fixed to the walls as shown in Plate 7.7. The wall is first coated with bitumen paint, applied with a long-handled soft brush. After about fifteen minutes it becomes sticky, and the sheet of polythene can be applied, rather like hanging wall-paper. Even better adhesion results if the wall has been covered with an undercoat of bitumen paint applied some weeks earlier.

300-gauge polythene sheet is preferred, as thicker sheets are too stiff and heavy and tend to peel off the wall before the bitumen is dry. For most purposes a 1.8 m wide sheet is placed lengthwise along the wall, with half the width of the sheet stuck to the top 0.9 m of the wall and the remaining half tucked away on top of the wall. This is pulled across the 'shoulder' when the silo is finally filled, a coat of mastic put on the upper surface, and a top sheet then put in position and sealed by pressing the two sheets together (Figure 7.1) and holding down with covering material (page 125). Additional protection can be got by laying the polythene sheet, even though possibly damaged, from the previous year's silo on top of the new sheet.

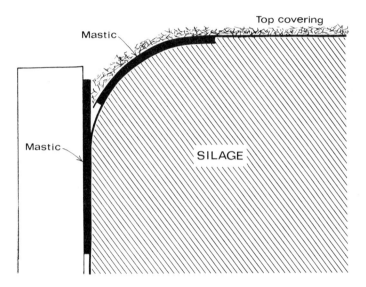

Figure 7.1   The use of side and top sheets to prevent wastage on the 'shoulder' of a bunker silo

As the silage settles the weight of material on top of the side-sheet peels it from the upper part of the wall; but the silage still presses against the bottom 0.5 m or so of the sheet against the wall, so that no air can enter. With this method of sealing the wastage commonly found at the 'shoulders' of silage made in walled silos can be almost completely prevented. The same procedure obtains for end walls, including any put in position before filling begins but designed to be removed to allow the silage to be fed out.

The Filling Procedure
To prevent oxidation and heating of the cut crop, the aim is to concentrate the crop in one part of the silo at a time. Thus the first loads that come in are dumped directly against the back wall of the silo, and

packed close beside each other. A fill of at least 1 m each day should be aimed at so as to give consolidation and act as a 'blanket' on previously-filled material. The need to ensure this minimum depth of fill will determine the area covered; thus in a silo of small dimensions it may be possible to fill to 1 m over the whole floor area of the silo on the first day. But in most cases only part of the floor will be covered, and the silo is then loaded in the form of a wedge, with a sloping face, up which the tractor runs.

The first day's fill will then have the wedge shape shown in Figure 7.2 (A). The slope established will depend on the width of the silo and the speed with which the cut crop is coming into store. As far as possible the 20° slope shown should be aimed at, but slow filling of a wide silo may require a rather steeper slope. A key feature is that the slope must always be shallow enough for the tractor to run up and down in complete safety.

As soon as the last load is spread in position each evening the exposed surface of the crop is immediately covered with a polythene sheet *to prevent air movement and heating during the night.*

This sheet is removed before the second, and subsequent, day's filling is added to the silo. This is done as in Figures 7.2 (B) and 7.2 (C), in daily layers of not less than 1 m of material, still maintaining the original slope, and progressing towards the open end of the silo. When the end is reached the top surface is levelled off to give a uniform finish, so that the whole of the material in the silo is at the required depth. Filling and levelling can be carried out with a variety of equipment; what is important is to adhere to the procedure just described, and to ensure that any one section of the silo is filled from floor to finished height *within three days.*

This method has particular advantage in allowing flexibility in organisation—it should not be thought that the standards indicated here can be achieved only by a large labour force and lots of machines. In practice the Dorset Wedge system of filling is applied very successfully on farms where *one* man does the whole job of cutting and filling. Equally it has proved effective in the hands of larger operators, including contractors. The key to success is in varying the filling slope, in relation to the harvesting rate and the width of the silo, so that the fermentation process is kept under control under a wide range of conditions.

Equipment for Filling
The cut crop is generally loaded into the silo by a tractor and buckrake; push-off buckrakes, either front or rear mounted, are preferred. Rear-mounted machines have the advantage that they can better mount the slope of the silo, but looking backwards can be a strain on the driver. Alternatively a high-lift front-mounted buckrake can be used (Plate

MAKING SILAGE IN BUNKERS AND CLAMPS

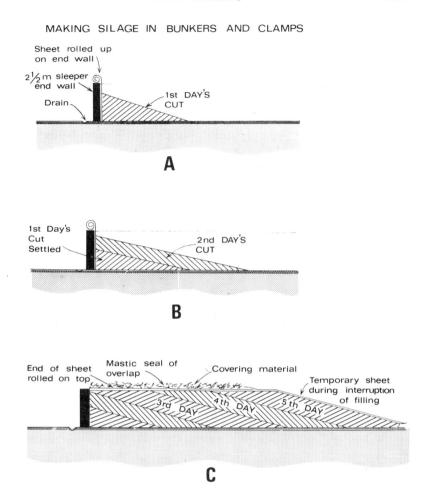

Figure 7.2   A–C. Stages in filling a bunker silo by the Dorset Wedge system

7.8). This type of machine has a working rate up to 40 tonnes per hour; the high lift and discharge height allows accurate placing of the loads of grass, but there is less immediate consolidation than when a buckrake is used, and more care then needs to be taken in covering the surface each night to restrict air movement through the more loosely-packed crop.

Special-purpose handling vehicles, now available on many farms, are well suited to loading silos, with their large-capacity forks, swift-acting hydraulics, and manoeuvrability. Four-wheel drive is a distinct advantage if full potential output is to be realised.

Whichever equipment is used for loading, there is advantage in filling

Plate 7.8   A front-mounted push-off buckrake for silo loading allows a high rate of work, without the need for the tractor to run on to the silo.

the silo progressively on a wedge, as this allows the second element of the system to be carried out—namely the final covering and sealing of the silage surface with plastic as soon as the finished height is reached. The aim is to fill each section of the silo to about 0.3 m above the likely final settled height within a period of three days, so that final covering proceeds from one end of the silo to the other as it is filled (Figure 7.2). As already noted, the exposed surface of the crop should always be covered with sheeting whenever there is a break in filling, both overnight and at weekends; this is done with a slightly-weighted plastic sheet that can easily be removed.

The passage of the buckrake tractor as it loads crop into the silo will give some consolidation, but if the chop length of the crop bears the right relationship to its dry-matter content (page 108), the weight of the crop itself will rapidly lead to a considerable degree of settlement, and this speeds up as soon as the crop begins to ferment. Some rolling may be necessary during filling, particularly if the crop is rather dry, but in general farmers spend far too much time on this job, and in many cases the constant passage of a tractor over the surface can produce a 'bellows' effect which actually assists the entry of fresh air. On no account should rolling be allowed to delay the final covering with the polythene sheet, since heating and wastage are certain to develop if sealing is not rapidly completed.

As soon as each section of the silo is filled to the required height the appropriate length of side-sheeting is pulled over the shoulder, mastic applied, and the top sheet laid on it and pressed into contact (Plate 7.9). It is then most important to apply some weighty material over the whole surface of this sheet so that it is immediately pressed into contact with the crop below to prevent entry of air. A 100 mm layer of chopped grass will do this, and is particularly recommended if straw bales are to be loaded on top of the silage later, since the grass will prevent the sharp ends of the straw damaging the sheet. Other suitable materials are farmyard manure (for outdoor silos), pit belting and old tyres; the latter are most effective, but must be used in sufficient numbers so that the whole area of the sheet is covered from view (Plate 7.10); this is most important with outdoor silos, where any exposed area of plastic will be weakened by sunlight and perforated by the claws of birds.

Outdoor Silos

It is a reasonable assumption that, if a satisfactory job can be done in protecting silage indoors from the damaging effects of air movement, then similar methods can be used to give weather protection as well for outdoor unroofed silos. The main difference is that it is advisable to use thicker polythene sheet, of at least 500 gauge thickness, or butyl rubber sheeting, to give the extra protection needed outdoors.

Much the same filling and covering procedure is used for outdoor *walled* silos as for indoor ones, but there is special need to ensure a weather-proof finish and to prevent rain running down the insides of the walls into the made silage.

Figure 7.3 shows an effective method of using polythene sheets on the

Plate 7.9 As each section of the silo is filled, the side-sheeting (Plate 7.7) is pulled across the shoulder and the top sheet is stuck to it, using mastic.

side walls; when filling is completed the sheets from the two walls are drawn across the surface of the silage to overlap at the highest point, where a bitumen seal is applied to keep out air and rain, before the top is weighted down with covering material. Any rain which reaches the surface of the top sheet will run towards the top of the walls, where it then flows down between the sheet and the wall and is carried away by the field pipe at the base of the wall. Great care must be taken in dealing with this effluent to prevent pollution.

A simpler form of silo can be made by constructing dwarf earth walls down the side and along one end of a concrete pad; it is often possible to take advantage of local topography by excavating into a slope. Sealing is effected by lining the walls with polythene sheet down to the concrete and overlapping this with a top sheet, weighted in position.

Plate 7.10 Outdoor clamp silo carefully sealed with polythene and tightly covered with tyres to keep the sheet in close contact with the silage to prevent surface wastage.

OUTDOOR WALLED SILO-END VIEW

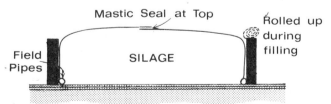

Figure 7.3 Sealing an outdoor silo to prevent entry of rain-water

Clamp Silos

The ultimate in economy and flexibility for storage is the outdoor silo without walls, since this can be placed at any convenient point on the farm where there is a level site on clear ground. If self-feeding is to take place a concrete base will be needed, but in other cases, unless the soil is very heavy, satisfactory results can be achieved by making the silo directly on the field surface.

Filling a clamp silo follows exactly the same principles already described for indoor silos, but some simple adaptation is necessary to compensate for the absence of walls, and to ensure that the covering sheet is pressed firmly against the silage so as to prevent wastage. The silo sides should have a slope no steeper than 20 degrees; the position at the end of the first day's loading shows how both the end slope and the filling slope are established (Figure 7.4 (A)). The second and subsequent days' filling (Figure 7.4 (B)) proceed with the aim of adding a minimum of one metre extra depth of crop each day, at the same time maintaining the shallow slope on the sides. As soon as enough length has been filled, and preferably after no more than two days, a 500 gauge polythene sheet is applied laterally across the silo (Plate 7.11) and

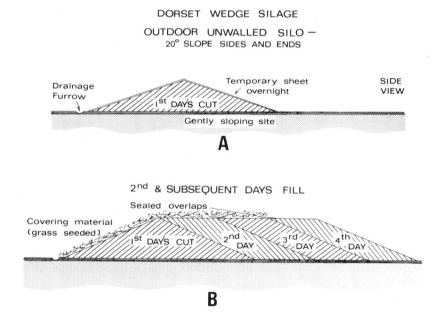

Figure 7.4   A and B. Stages in filling an outdoor sealed wedge silo

Plate 7.11   500-gauge polythene sheet being placed in position across the end of an outdoor wedge clamp silo immediately after loading.

immediately covered. As noted, old tyres are very effective for this purpose but other covering materials, including FYM or soil can be used. If FYM is used the main application can be made by driving alongside the silo with a rotary spreader. Whichever material is used, the whole surface of the sheet must be covered, and the layer should be at least 150 mm deep to allow for the washing effect of rainfall; it is here that the shallow slope to the sides is so important. The covering material can also be more firmly anchored by making a light sowing of grass or cereal seeds, which will root into the manure, or soil, and bind the whole together against washing by rain. This can also make removal of the covering much easier, since the whole mass will be rooted and can be pulled off in a mat.

Prompt sheeting and weighting of the cover will go a long way to preventing surface wastage, even in the case of the sloping sides which will have had very little consolidation. But any delay in sheeting and sealing will lead to heating and oxidation of nutrients, and subsequently to surface wastage, which will render the whole operation a waste of time.

It must be stressed that great care must be taken by the tractor driver in building this type of outdoor clamp, although the risk of overturning is much less than with the conventional type of clamp with vertical sides, as it is not necessary to consolidate the sloping sides to prevent over-heating, as long as they are properly sealed.

To prevent waterlogging around the silo, particularly on heavy soil, a shallow trench should be ploughed around the site to direct the run-off to a soakaway or to a pit that can be pumped out.

Ensiling in a Polythene Sleeve

Another technique which avoids the need for a fixed retaining wall is to enclose forage in a large-diameter polythene sleeve, using specialised equipment. Short-chopped forage is pressed into the sleeve, which has one closed end, by a slow-speed rotor supplied with crop from a feed table. Extension of the sleeve is resisted by a retaining mesh controlled hydraulically by a cable and brake so that the forage is evenly and densely packed. A filled sleeve is typically 2.7 m diameter and 40 m long, containing about 130 tonnes of silage, depending on the density and moisture content of the grass.

The packing mechanism can be powered either by tractor or by integral diesel engine, when the claimed work-rate is 30–50 tonnes per hour.

Quick and effective sealing is an inherent and highly desirable feature of the system. The double-skinned polythene is kept taut by its contents if properly filled; however, it remains vulnerable to damage and should be fenced off from animals and inspected regularly so that any punctures are soon repaired.

Silage is usually unloaded from the sleeve with a tractor foreloader and grab; a firm working base is essential for this operation. With its high capital cost the system appears most suited to contractor use, for the loading unit is readily moved from farm to farm and is well adapted to making relatively small lots of silage, particularly where existing storage facilities are inadequate; it could have particular application for ensiling small areas of forage maize.

Big-Bale Silage

The attempt to make silage by sealing individual bales within polythene bags was first made with 'standard' bales, but the laborious man-handling involved prevented application on any practical scale. However, the introduction of the large roll bale with fully mechanised handling has led to rapid development of this system.

The technique of ensiling large bales has a number of useful features. Small quantities of forage can be ensiled at any one time, and because a silo construction with retaining walls is not necessary the capital required for storage is minimal. Baling can also be carried out independently of the transport and storage operations, which makes it well suited to contractor use.

While the initial development was with balers primarily designed for hay, balers have now been developed specifically for baling green crops. These have a relatively low pto power of about 30 kW, although a

Plate 7.12 Roll bales of silage enclosed in individual polythene bags and stacked for storage. The stack is covered with a net, weighted to minimise wind damage, and fenced against stock.

tractor with twice this rating may be needed if the unit is required to operate on slopes. Roll bales 1.2 m × 1.2 m diameter, each weighing 300–500 kg, are the most common (Plate 7.12). However, one machine produces bales of a similar weight, but rectangular in cross-section (1.6 m × 1.2 m × 0.7 m) and tied with four substantial galvanised wires. These bales stack better than round bales, but this machine is more complex, and more expensive to purchase and operate, than a round baler.

As with all silage-making systems, airtight conditions must be quickly established and maintained; this is particularly important with silage made in separate packages because of their relatively low density. Thus the most common practice is to place each bale into an individual polythene sack just prior to stacking; during stacking some of the air in the bag is expelled, and the neck is then tightly closed. It is now recognised that it is essential to use high-grade plastic for the bags; in addition special care needs to be taken to protect the bags from damage by wind or animals during the storage period. Particular attention must be given to preventing rodent damage; this is best done by baiting the approaches to the stack of bales, rather than the stack itself. There is also advantage in covering the whole stack with a net, weighted to prevent wind damage. With such precautions excellent silage should be made with minimal loss.

Wrapping Big-Bale Silage

Although effective, placing individual bales in bags and tying them is a tedious operation. Techniques have therefore been developed to wrap round bales of cut forage in a tough stretchable cling-film. The bales are tightly wrapped and should store well with little wind damage. Once wrapped the individual bales are currently handled with a conventional spiked loader, which requires the holes to be resealed.

Three basic systems of applying the plastic are being used. With one machine plastic is wrapped as the bale is rotated in two directions on a revolving turntable (Plate 7.13); another gently rotates the bale along the ground as a roll of film spins around it (Plate 7.14); in the simplest system plastic is applied from a hand-held dispenser as the bale is rotated on a rear-mounted prong on the tractor. At the time of writing there has been insufficient practical experience with any of these systems to assess their effectiveness. However, they do have considerable appeal, and it is likely that efficient machines will be developed which could be well suited in particular to contractors and larger operators.

Some operators have also successfully enclosed a stack of unbagged roll-bales within a carefully sealed double covering of polythene sheet. However, only as many bales should be sealed within one pack as will be fed out during one week, because bales exposed for any longer period are likely to deteriorate badly. A similar method has been used to ensile stacks of rectangular bales.

If possible the forage should be wilted to 30 per cent dry matter or above before it is baled, as fermentation of very wet baled crops can be poor, and it is difficult to apply a silage additive uniformly during baling. Wilting also makes the bales lighter for handling, and fewer plastic bags are needed for each tonne of crop dry matter stored.

Big-bale silage has been an important innovation in introducing silage to farmers who did not consider their operation large enough to justify

Plate 7.13   Wrapping a roll bale with 'stretch' film, using a rotating turntable system (Photograph Dairy Farmer).

Plate 7.14   Wrapping a bale with film by rolling it along the ground inside a rotating film dispenser.

investing in conventional harvesting equipment; it has also proved useful in conserving smaller lots of silage on farms which have already conserved large amounts of silage in the main growth period in spring and summer, so avoiding the need to open up an already sealed silo. Baled silage is well adapted to mechanised feeding out to stock (Chapter 9) and has the particular advantage that it makes it easy to feed silage to grazing stock (page 160) at times when there is a shortage of herbage for grazing, without the need to open up a main silo. Baled silage is also convenient for off-farm sales.

However, the overall work-rate can be limited by the high labour requirement in sealing and storing the bags, and big-bale silage is most unlikely to replace the well-organised forage harvester system on the larger farm. Again it must be stressed that success with the system, perhaps even more than with other silage systems, depends on meticulous attention to detail.

OTHER ASPECTS OF SILAGE

Ensiling Forage Maize

As the maize plant matures its moisture content falls, and this is likely to be the main factor determining the date at which it is cut. Below 20 per cent DM there is considerable effluent loss from the silo, and for ensiling in bunker or clamp silos a dry-matter content in the range 24–28 per cent should be aimed for. However, in a late season, to delay cutting until this level is reached may mean the soil becoming too wet to take the harvesting equipment, and there may sometimes have to be a compromise between this and the 'fitness' of the crop.

The harvested maize can be loaded into the silo with a buckrake or a foreloader and its friable nature allows very high work-rates. Quick filling and prompt and efficient sealing are even more important than with grass crops, for if air gains access to the heap of maize, heating is very rapid. Particular care must be taken with sealing down the covering sheet if the maize is to be stored in an outdoor silo.

Because of its high fermentable carbohydrate content, a chemical additive is generally not needed for ensiling maize. However, and particularly at higher dry-matter contents, a problem may arise with secondary fermentation and deterioration of maize silage when the silo is opened. Experiments have shown that this can be largely prevented if eleven litres of propionic acid is added per tonne of crop at the time of ensiling. But this is expensive, and it is better to chop the maize finely to speed consolidation and so prevent air getting into the silage, both during the making process and also when the silo is opened.

The low protein content of maize silage has already been noted, and research has shown that it can be very effectively supplemented with a source of non-protein nitrogen, such as urea. This can be added to the silage just before it is fed; alternatively a proprietary 'protein' product, containing urea or ammonium salts, can be added at the time the maize is ensiled. To ensure complete mixing this is best done, as with silage additives, during the harvesting process.

The Use of Silage Additives

There has been a big increase in the use of silage additives during the last decade, and in the range of chemicals used (page 22). Several different methods of application are also available, suitable for both solid and liquid additives; liquids are now generally used because they are easier to handle and do not block in wet weather. Where only a few hundred tonnes of silage are to be made, equipment based on a 25-litre container of chemical feeding directly through a tube into the flail or chopping mechanism is satisfactory (Plate 2.1). It is then important that the feed pipe is inserted at a point on the forage harvester where air suction will draw the liquid from the nozzle and break it into droplets. Some manufacturers of both harvesters and applicators have failed to identify this position correctly, and the fitter should experiment with a simple smoke generator to get the best location. This will also avoid positions where back-pressure may cause spray and fumes to escape.

To avoid handling a large number of small containers it is now common for either the tractor or the harvester to be fitted with a cradle to carry a 200-litre drum of additive (Plate 7.15), generally feeding in to the cutting mechanism via a simple glandless pump, controlled either by the driver from the tractor cab or by an automatic cut-off device on the pick-up.

The amount of each particular additive applied should be as recommended by the manufacturer, taking note of all the crop and weather factors summarised in Table 2.3. Rate of application is generally controlled by the size of the nozzle used at the end of the delivery tube into the cutting mechanism, or by varying the speed of the pump, when one is used.

Some additives, particularly those containing acids, are corrosive, and can damage eyes and skin. In handling them special care must be taken to follow the manufacturer's instructions, particularly with regard to the use of gloves and goggles during filling and changing containers. It is also important to cut off the flow of chemical whenever crop is not passing through the harvester, as on field headlands, so as to avoid spray drift.

Plate 7.15   *Silage additive is best supplied from a 200-litre drum, so as to minimise delays from frequent handling of containers.*

Accurate and uniform application of the liquid to the crop is of crucial importance to the successful use of an additive. Thus it is a real advantage to explain to the harvester operators why the additive is being used and how it works. They are then much more likely to apply the correct rate, and to change this at intervals as crop condition changes. If possible a number of trailer-loads of crop should also be weighed, for few farmers really know how much their trailers hold—and without this information selection of the correct application rate is very much a matter of guesswork.

Silage with Reduced Fermentation

The possibility was noted in Chapter 2 of silage preserved partly by sterilisation, rather than by acid fermentation. Several mixtures of acids with formalin are available which restrict fermentation. They produce a rather different type of silage, in both appearance and smell, but the procedure for making the silage is identical to the methods already described, with rapid filling of the silo in regular daily increments followed by prompt application of polythene sheets. In fact care during storage and feeding out may be even more important than with conventional silage, since the lower content of fermentation acids makes the silage more vulnerable to spoilage organisms and oxidation if air is allowed to get in. A noticeable feature of the products, including those made with formaldehyde, is the absence of a typical 'silage' smell, possibly important with the greater emphasis now put on the environment. There is also some evidence that less effluent is produced because of the reduced fermentation in the silo.

The silage produced with the use of an effective additive should not only be preserved better and with lower losses, but it should also be of higher nutritive potential. Advantage must be taken of this so as to justify the extra costs involved—though it is worth noting that the increase in silage intake resulting from use of an additive could mean that more silage has to be made. This may well modify the way the farmer approaches the planning of his crop conservation, with the conclusion that it is worth taking considerable care both in making and feeding silage.

Effluent Loss from Clamp and Bunker Silos

A compacted wet crop loses moisture, which runs out of the silo as effluent. The loss of dry matter contained in this effluent is seldom more than 1 per cent of the crop, but it is a valuable fraction, containing readily digestible soluble components and minerals. More seriously, effluent creates the social problem of smell and the legal problem of the pollution of water-courses. So every effort must be made to limit effluent—and when it *is* produced, to trap it so that it does the minimum of damage.

More than 10 per cent of the fresh weight of the crop—up to a hundred litres per tonne—can be lost as effluent from a crop loaded into a bunker silo at below 20 per cent DM content; but effluent is virtually eliminated if the crop has been wilted to 25 per cent DM. Hence the general advice is to wilt the crop whenever possible; this is still the most effective way of reducing effluent.

But under adverse weather conditions it is often necessary to load crop at under 20 per cent DM, at which level liquid can readily be squeezed by hand from the cut crop. It is then important to avoid heavy consolidation, and to prevent heating by careful sealing, for heating and oxidation of sugars produce water which can further increase the moisture content of the crop. The effect of additives on effluent flow may vary; formic acid seems to speed up the first flow, but may produce no more total effluent than untreated silage, while additives containing formaldehyde appear to reduce effluent slightly.

Where effluent is produced it must be trapped by half-round drains laid across the fall of the floor of the silo, and led away to a catch-pit. From this it should be diluted and sprayed on to the fields where the crop was cut; if this is done daily, before the effluent begins to decompose badly, 2,000 litres or more can be applied per hectare without damage.

Recent work at Hillsborough, N. Ireland, has also shown that fresh silage effluent can effectively be fed to beef cattle. Yearling cattle readily consume up to 14 litres of effluent per day, and the 0.65 kg of dry matter contained in the effluent has about the same feeding value as the same weight of barley.

Tower Silos

We have given particular attention to the use of surface silos because we believe that this will be the method of storage adopted by most silage-makers, at least for the foreseeable future. In the late 1960s, however, many British farmers saw tower silos in use in the United States, where they were proving especially valuable for ensiling lucerne and whole-crop maize (corn) as part of a fully mechanised system of livestock feeding. As a result a considerable number of silage towers were installed in the United Kingdom, in an attempt to capitalise on the efficiency which a sealed container and full mechanisation can contribute. This development tended to be localised in certain areas, such as the south-west coastal area of Scotland, where prevailing weather conditions had proved favourable, or where the enthusiasm of a successful operator had proved infectious.

A good deal of uninformed controversy developed between the protagonists of clamp silage on the one hand and of tower silage on the other. We do not subscribe to the view that they represent two quite different concepts of silage-making. Indeed the more one examines the principles and practice of making tower silage, the more evident are the similarities with the standards and the recommendations already made for clamp and bunker silage.

By the very nature of their design tower silos ensure exclusion of air from all but the top layer of silage, and so offer very efficient storage. However, to get full mechanisation of filling and feeding out demands more precise control of both the dry-matter content and the chop length of the crop than is needed for modern surface silo techniques. Forage with a DM content of 35–40 per cent is preferred as this exerts less pressure on the silo walls than a wetter crop, and also avoids production of effluent, which is difficult to manage with towers. Crop at this dry matter is also handled more satisfactorily by the tower unloading equipment, the reliability of which has been greatly improved as a result of more robust construction—and the recognition by operators of the need for regular maintenance. For the same reason, and also to aid consolidation, the forage must be short chopped, with a median length of 35 mm, and certainly free from pieces longer than 100 mm.

More recent top-unloading equipment has been designed to tolerate some variation in chop length. It employs a rotating finger-wheel rake mechanism to distribute the crop during filling, and then to collect the silage into the intake of a pneumatic conveying system during emptying. The electrical requirement of this equipment can be as low as 3 kW, an advantage on an isolated site if a tractor or diesel engine can be used to operate the main conveying fan for filling the tower. Tower silo installations require careful planning to take account of such factors as siting to provide easy access of crop for loading and for feeding out, as well as possible scope for expansion. The dimensions of the tower must also be planned so that at least 100–150 mm of made silage is removed daily during unloading so as to prevent surface spoilage. The tower should also be sited so as to have minimum effect on the landscape.

The main consideration, however, should be in the crops to be ensiled. To justify the high cost of the tower installation the silage that is fed out must be of high nutritive value; thus the crop loaded into the tower *must* be of high digestibility. As discussed in Chapter 3 this means cutting at an immature stage of growth, which poses the practical problem of wilting such crops, often cut early in the season, to the 35 per cent dry-matter content needed for the tower. Application of the latest methods of cutting and field wilting, described in Chapter 5, can help greatly in getting crops cut and wilted at the right stage.

Clearly a tower silo is an expensive investment; but it offers a very efficient method of conserving forage, linked into a fully mechanised feeding system. Thus it could certainly be one of the options examined when a new forage storage and feeding operation is to be installed.

Care During Storage of Silage

Six months or more are likely to elapse between sealing the silo and the time when it is opened for feeding out, and during this period it is important to examine the silo at intervals. Wind and rain may cause areas of the sealing sheet to become exposed, and these should be covered over immediately so as to keep the sheet in close contact with the silage underneath.

Outdoor clamps should be fenced off, since it is common to find they have been walked over by cattle, sheep or even deer, with disastrous results to the sheets—and to the silage. The silo should also be checked at intervals to see that no rain is getting in, and that there is no faulty drainage of the site, which can lead to the lower layers of silage becoming waterlogged.

At the time of covering the silo provision should be made for sampling the silage by marking with a piece of distinctively coloured material, such as a fertiliser sack, two or three suitable points where the sheets overlap. This will avoid breaking the air-seal more than is absolutely necessary when a core-sampler is later used to sample the silage. The advisory services will sample every type of silage, and their nutrition laboratories will provide farmers with analytical details, and with feeding recommendations for the particular stock to be fed.

Considerable care is needed if retaining walls have to be removed to expose the silage for feeding, because of the pressure the silage may be exerting. As the external walls are taken away the polythene lining sheets should be revealed still firmly adhering to the silage and, if the making and storing suggestions described here have been successfully followed, the silage below should be completely free from surface wastage. The days when the farmer could spend the best part of a day hauling away rotted material before he reached edible silage must surely be a thing of the past.

Some of the methods we have described for making bunker and clamp silage may seem elementary, but their aim is to produce an entirely predictable material free from surface wastage. For it is only necessary to make the most cursory check on current silage-making operations to see how often serious surface wastage still occurs—and with it the losses associated with overheating and effluent loss from the silo. These represent one of the most serious misuses and depletions of expended resources in the whole range of agricultural activities; yet in many cases only quite minor changes in methods, and in their timing, can promote a dramatic improvement in results. This is within the grasp of every silage maker.

The more sophisticated and costly harvesting and handling procedures clearly need to be examined critically to ensure, before they are adopted, that they are likely to contribute specific economic advantage. In particular, if a higher-value product is being made, it is important to calculate the cost of achieving this, as against the cost at which alternative feeds could give the same level of animal output. In our view such alternative feeds are likely to become more expensive. In that event, conservation systems that are more efficient, and in some cases more expensive, will be fully justified.

Chapter 8

STRAW AS ANIMAL FEED

CEREAL STRAW has always been used in animal feeding. But straw has contributed little to the rations of productive livestock because it is very poorly digested by ruminants—D-values are generally below 50—and its intake is also low. Straw also has a very low protein content.

Straw is indigestible because the fibre it contains is firmly bound together with lignin, needed to give the straw enough strength to hold up heavy ears of ripening grain. But this lignin prevents the rumen organisms getting into contact with the fibre when the straw is eaten. It has been known for many years that when straw is treated with alkali some of the lignin is dissolved away and the fibres also swell and become more open; both these changes make it easier for rumen organisms to digest the fibre. Interestingly (as was the case with silage additives) the process, in which straw was soaked in caustic soda solution and excess alkali washed out before feeding, was first used in Scandinavia, though with only limited success. The high animal feed prices of the 1970s led to renewed interest in the use of treated straw. Several processes, which differ in a number of ways from the earlier processes, have been developed. In general they are more mechanised, and so require less hand labour; this has made it possible to use more concentrated alkalis, and in some cases heat, which has made delignification of the straw faster and more complete. Excess alkali is also now not washed out of the straw, because it is recognised that it causes no harm to stock, and may in fact be of some nutritional advantage.

The renewed interest in straw treatment coincided with more straw becoming available as a result of the huge increase in cereal production that was taking place. By the early 1980s annual straw production in the United Kingdom was well over 12 million tonnes, at least 7 million tonnes more than was needed for traditional farm uses for animal feeding and bedding, potato storage, etc. Yet only a small part of the surplus has been processed for animal feeding. The reason is that most of the surplus straw is being produced in the south and east of the

141

country, while most of the livestock that could use it are in the west and north—and straw is costly to transport over long distances. As a result much of the surplus straw is burnt in the field—burning is a faster way than baling and carting for clearing straw from the field in preparation for planting the next crop, and burning also helps to control weed seeds and cereal diseases. While increasing public concern about environmental damage from straw burning is leading to less straw being burnt, most farmers have responded by ploughing in the straw rather than baling it, because of the uncertain market for baled straw.

However, large areas of cereals *are* grown within reasonable distance of major livestock regions, and with the pressure, particularly on dairy farmers following milk quotas, to reduce feed costs more interest is now being shown in the use of processed straw for ruminant feeding. Straw will mainly be fed as a part replacement for conserved forage—hence the present chapter.

The Digestibility of Straw

Straw can differ widely in feeding value, depending on the cereal grown, the variety, the presence of weeds in the straw (some weeds are more digestible than the straw itself!) and how much the straw has been weathered before it was picked up. Spring barley and oat straws are generally more digestible than those from winter barley and winter wheat—possibly reflecting the stronger straw needed to hold up the heavier winter crop. Few lots of straw though are more digestible than even badly made hay. But after treatment with alkali, the D-value of all types of straw can be increased to the 55–65 range, as high as most of the hay and some of the silage now made. Perhaps unexpectedly, alkali treatment increases the digestibility of low D-value straw more than that of high D-value, so that treated straw tends to vary less in digestibility than untreated straw.

Collecting and Storing Straw

The aim of the many investigations carried out since the early 1970s has been to develop practical methods of treating straw consistently and economically. Straw is a bulky low-value commodity which must be delivered at low cost if it is to compete with alternative feeds. Clearly it is cheapest when it is used on the farm where it is grown, for transport can increase its real cost two or three times. Thus while for transport over short distances standard straw bales may be acceptable, over longer distances high-density bales are preferred, weighing over 200 kg/m³, for these nearly double the load of the average lorry. Few balers can currently produce such dense bales; those that do are expensive machines that must be used to full capacity to achieve the yearly output of 2,000 tonnes or so needed to make their use economical. These machines also have a high hourly output, which allows fields to be cleared quickly in preparation for the next crop. For the same reason, where transport distance is less than about 50 km, large round bales are now preferred to standard bales because they allow fields to be cleared faster.

Straw to be treated with alkali should be clean and between 15 and 20 per cent DM. Although large stacks can be stored in the open without much damage some form of protection, particularly from rain, is desirable. This is most simply done with plastic sheeting over the stack, held in position with a weighted net to prevent wind damage. Reinforced plastic sheets, though more expensive, give good protection and, with careful use, will last for several years. Interestingly, a simple 'pole-barn' structure, which gives secure storage and can also be used for other jobs, can be built for little more than twice the annual cost of plastic sheeting. However, sheets are needed if the straw is to be treated with ammonia, which requires the stack to be totally enclosed.

Treating Straw

Two chemicals, sodium hydroxide (caustic soda) and ammonia, are generally used for treating straw.

Sodium Hydroxide

This strong alkali is used in concentrated (27 per cent) solution in water, rather than in the more dilute solution of earlier methods. Added at the rate of 50 kg of sodium hydroxide (just under 200 litres of solution) per tonne of straw it gives rapid delignification. The problem in practice is in getting the solution uniformly mixed with the straw before it is put into store, for once in store there is little movement of the alkali so that any straw which is not wetted does not have its digestibility improved. Alkali is best applied as a spray to chopped straw, which is then mixed by paddles operating in a chamber before the straw is blown into store (Plate 8.1). In this process, which is generally contractor-operated, the material in store heats up to 70°C, at which temperature delignification is very fast, and is likely to be complete in three to four days. A 70 kW tractor is needed to operate the straw chopper and mixer, and output is about 2 tonnes per hour. Larger round and square bales can be treated in a tub-grinder in which the alkali solution is applied at the grinding rotor to aid uniform mixing. Output can be up to 4 tonnes per hour. The output of all machines is much reduced if the straw has been allowed to

Plate 8.1   Straw from small bales is chopped before being mixed with alkali by paddles in
an enclosed chamber. The treated straw is then blown into store
(Photograph K. A. McLean).

get damp before processing. In another system the alkali solution is
injected directly into large bales through a series of hollow tines inserted
into the bale. However, treatment is likely to be less uniform than with
chopped straw because the alkali is less evenly distributed. In all cases
the treated straw must be protected from rain because the straw tends to
mould very quickly if the alkali is washed out.

These on-farm methods are now being used on many farms which
have reasonable access to cereal straw. However, they generally give
rather less increase in D-value than would be indicated by the experimental results, with *in vivo* digestibility generally below 60 per cent,
because of the practical difficulty of mixing the alkali uniformly with the
straw. 27 per cent sodium hydroxide solution is also very corrosive and
must be handled and applied with great care—hence the advantage of
employing specialist operators. Treated straw also contains excess
alkali, and animals fed treated straw produce more urine than usual so
as to excrete the extra sodium ions. This may not matter with stock
housed in cubicles or on slats but it can mean more straw being needed
when animals are bedded in loose housing.

The first two of these problems have been overcome in an industrial-
scale process operated by one of the major animal-feed companies.

Baled straw is first broken down in a tub-grinder, and then hammer-milled. The milled straw is treated with a metered amount of sodium hydroxide and then thoroughly mixed before being extruded through a pelleting press. The combined effects of the high pressure and the temperature generated in the press give very rapid delignification of the straw; the pellets, after cooling, are also very suitable for storage and handling. Most of the treated straw pellets now being produced are used as an ingredient in dairy compound feed, in which their residual alkali content is useful in reducing rumen acidity when the compound is fed, and so in helping to maintain milk butterfat level (page 44). However, the major expansion of this process, which was predicted, has not happened, possibly because the straw pellets have shown little cost advantage over alternative feeds.

Ammonia

Ammonia is a weaker alkali than sodium hydroxide, and so delignifies straw more slowly; but it does have a number of advantages. As has been noted, to give effective treatment sodium hydroxide must be intimately mixed with the straw. In contrast ammonia is a gas, which can *diffuse* through a stack of straw so that complete mixing with the straw before it goes into store is not essential; many farms are also already equipped, and with a national supply service, to use anhydrous ammonia as a nitrogen fertiliser; and some of the ammonia reacts chemically with the straw to produce compounds which can provide a source of non-protein nitrogen which may reduce the need for supplementary protein when the straw is fed (page 43).

Ammonia can be applied either as anhydrous ammonia (a liquid under pressure) or as a 35 per cent saturated solution in water. In both cases the straw must be totally enclosed in some type of sealed container to prevent loss of ammonia to free air. This is generally done by enclosing a stack of baled straw within plastic sheets. Standard bales are typically built into a stack fifty bales long by four or five bales wide and seven or eight high, and holding some 30 tonnes of straw. The stack is built on a plastic sheet laid on the ground (taking care to avoid damage from stones or flints), with a slight 'ridge' to shed water, and covered with a further sheet which is 'sealed' all round to the ground sheet by weighting, generally with spare bales. The sheet must be held securely against the stack by a net or ropes to prevent it flapping in the wind, which would pump ammonia out through any small cracks. The required amount of ammonia (calculated at 30 kg of ammonia per tonne of straw in the stack) is then injected through a hollow probe inserted through the plastic at points round the stack—remembering to seal the holes afterwards. An interesting innovation has been to join the plastic sheets

Plate 8.2 Aqueous ammonia is injected into straw, either in large bales or in a stack, enclosed in polythene sheet.

with double-tubing seals similar to those used in the New Zealand vacuum silage method popular in the 1960s.

Large roll bales can be stacked and treated with ammonia in the same way (Plate 8.2), though particular care may have to be taken in securing the covering sheet because the stack is likely to be less rigid than one made with rectangular bales. Alternatively large bales can be treated, either singly or in groups of three or four, in a long plastic tube. This system takes more ground area than a stack, and the site must be carefully selected to allow access in winter (Plate 8.3). In another system anhydrous ammonia is injected into roll bales through five hollow tines fitted to a tractor fore-end loader, and supplied with ammonia from a tank on the three-point linkage. The treated bales are then pushed through a collar into a long plastic tube, which unfolds from the collar as further bales are inserted. Up to fifty-five bales can be loaded in this way into a single tube. As with all forms of storage it is essential for the plastic to be held firmly against the bales to prevent wind damage and loss of ammonia.

Another approach is to line the walls of an existing building with plastic sheeting before loading the bales. A top sheet is then sealed to the top of the wall sheets before ammonia is injected. This gives good protection for the plastic as well as good working conditions for feeding out in winter.

Plate 8.3 *Large roll bales inserted in a polythene sleeve before aqueous ammonia is injected. These occupy considerable ground area, and the site should allow easy and firm access for feeding out.*

As already noted, ammonia delignifies straw more slowly than sodium hydroxide, and straw generally needs several weeks of treatment before it is ready for feeding—possibly as long as eight weeks during cold weather. Anhydrous ammonia also has only limited effect if it is applied to very dry straw, containing less than about 15 per cent moisture. On the other hand, if it is applied to very wet straw it tends to be absorbed by the moisture close to the injection points; this can give non-uniform treatment, and the heat generated by the chemical reaction can also make some moisture migrate and form spoiled wet patches farther away. Overall the optimum moisture content in straw to be treated appears to be about 20 per cent.

Although ammonia treatment of straw has given good results under experimental conditions, practical experience with on-farm treatment has been less successful. Thus in a study on 153 farms, Ibbotson, Mansbridge and Adamson of ADAS found an average increase in D-value of only 10, from 44 to 54, compared with 60 plus in many experiments. Much of this lack of success is likely to have resulted from the difficulty of achieving uniform distribution of the ammonia in large-scale farm treatment. It also appears possible that some farm stacks of damp straw may have been left covered for several days before the ammonia was injected; under these conditions carbon dioxide could have been produced by fermentation of the straw and neutralised some of the ammonia, so that insufficient free ammonia was left to delignify the straw completely. Clearly much care is needed if ammonia treatment is to be effective in practice. And, where sufficient ammonia has been applied, it is important to open the stack several days before the straw is to be fed, so as to allow any excess ammonia gas to escape.

Oven Methods

Ammonia reacts faster with straw at higher temperatures, and several systems have been developed in which straw is treated with anhydrous ammonia in a heated, sealed oven. Ovens hold between 1 and 5 tonnes of straw, and ammonia gas is recycled through the straw, which is heated to 90–100°C (Plate 8.4). Delignification is complete in about twenty hours, and this is followed by a period during which the oven is vented to atmosphere to get rid of any surplus ammonia before the straw is unloaded. It has been suggested that ammonia may be less efficiently used in ovens than in the stack method, and in one oven system the final exhaust gases are vented into bales enclosed in a plastic sleeve so as to utilise any residual ammonia.

Loading and unloading the oven is likely to require some hand labour, but supervision of the treatment process is largely automatic,

Plate 8.4   Bales of all sizes can be treated in an 'oven' with anhydrous ammonia. The gas is heated and recycled to speed up the process during the twenty-four hour treatment cycle.

with timers and solenoid valves controlling the heating cycle and the injection of ammonia. Some of these control units have now been adapted to treat much larger amounts of straw, held in insulated containers (which can be constructed in existing buildings), but without using supplementary heat. Treatment takes longer than in a heated oven, but the heat of reaction can raise the temperature of the straw to 50 °C, and delignification can be completed in fourteen days. The saving on heat makes this a potentially attractive system, and several buildings in Scotland have now been adapted; lined with plywood and fitted with insulated doors they provide an effective way of treating lots of up to 20 tonnes of straw.

Chapter 9

METHODS OF FEEDING

THE MAIN OBJECTIVE in conserving forage must be to produce sufficient feed of suitable digestibility, intake and nutrient content for the particular animals to be fed. Methods of storing and feeding will depend much on the size and layout of the farm and on the number and type of stock; whichever method is used, the forage should be fed so that intake is not restricted below the planned level, at the same time avoiding physical wastage, contamination and deterioration of the feed. Much of the effort put into conserving forage *can* be wasted if the method of feeding is not carefully planned and executed.

HAY AND STRAW

Much hay is wasted during feeding, often because it is considered to be of too little value to warrant much care. Hay is often fed to outwintered stock, which only need a maintenance level of nutrition, by distributing bales over the field, with inevitable losses from contamination and treading. The alternative is to feed it from racks or feeders, but these tend to concentrate the effects of treading, which is bad for both soil structure and the stock, and also makes access by tractors difficult. Such disadvantages, however, are often accepted as a preferable alternative to providing a permanent hardcore base or a building, with the extra work and lack of flexibility which these can introduce.

However, most hay is probably fed to loose-housed stock; buildings are often out-dated and have difficult tractor access, and 'man-handleable' bales are then of great advantage. These are often placed in racks or troughs from which there can be much wastage by stock pulling out and dropping feed. Wastage can be reduced by lining the racks with weld-mesh to restrict the amount animals can pull out at one time. The selective feeding habit of sheep presents special problems which can lead to high wastage; a welded mesh or similar grid placed on top of the trough can limit waste.

Similar principles apply to feeding hay from big bales. As noted, large rectangular bales are made up of a series of bundles, each weighing approximately 10 kg, which assist feeding within a confined building. The remainder of the bale can be retied to allow it to be transported to the next feeding point. Whole bales are best fed from within a rectangular cage feeder with weld-mesh sides, which match the dimensions of the bales and are designed to be pressed against the face of the bale as it is eaten. Both housed and outdoor stock can be fed from large roll bales, placed by foreloader into a feeder which matches the size of the bale and has all-round access for stock. Alternatively, round bales can be placed behind a feed barrier along a feed passage. It will then probably be necessary to move the remains of each bale forward within reach of the stock as feeding progresses. Round bales can also be unrolled to leave a band of hay or straw on the ground. Simple machines have been developed for this purpose, but some tidying up by hand may be needed to avoid wastage.

It is not possible to specify the optimum method of feeding, because this depends on the type of stock to be fed as well as on building shape, layout, and access. Design is generally a case for local ingenuity, aiming to ease labour requirement and to minimise waste without limiting the amount of feed the animals can eat.

Loose hay, dried in bulk, can be extracted from the store with a tractor foreloader and grab, which effectively contains it for transport over short distances and then places it in racks lined with mesh. Where animals are to be fed at some distance from the store feed racks can be mounted on wheels and also used for transport. Loose, partly chopped, hay is in an ideal form to be mixed with other ration components in a mixer wagon. Blocks of hay 2.0 m × 1.0 m × 0.9 m and weighing 200 kg, can also be cut from the store with specially designed tractor-mounted equipment.

SILAGE

Most silage in the United Kingdom is made in walled bunkers or unwalled clamps. From these it is either fed directly from the face (self-feed), with the top layers from high silage faces thrown down by hand (easy-feed), or extracted and transported for feeding at another site in some form of mechanised feeding.

Self-Feeding (Plate 9.1)

On many specialised livestock farms silage is now a major component of the winter diet and self-feeding is preferred because it is independent

Plate 9.1   Self-feeding silage systems avoid the need for complex machinery, but must be planned and operated with care to avoid wastage of silage, or restricting the amount of silage the stock are able to eat.

of machines and avoids the need for additional feeding arrangements, except for concentrates. Self-feeding systems do however require careful planning and operation to avoid the risk of restricting silage intake. This can arise from several causes, apart from intrinsic poor quality in the silage. Thus stock may find it difficult to pull long, heavily compacted silage from the face, a particular problem with young animals and for those with poor teeth. Space is also important. Where stock have 24-hour access the following face-widths are advised: 0.05 to 0.08 m per ewe; 0.15 m for young cattle; and up to 0.2 m for adult cattle and dairy cows, increasing to 0.7 m if all the stock need to have access to the face at the same time.

As well as ensuring adequate silage intake the method of feeding must also avoid physical wastage and deterioration of the silage from air getting into the face. An electrified wire or bar positioned across the face is the most common method of controlling the rate at which the silage is eaten back. The bar must be moved frequently, and the width of exposed face controlled so that it is eaten back at least 0.15 m per day so as to minimise ingress of air. Railway sleepers or telegraph poles placed at the foot of the face help to keep the silage behind the bar and limit losses by treading and contamination by slurry; slurry should also be scraped away regularly.

Self-feeding is most effective with short-chopped silage, 25–50 mm long, and uniform in density and quality throughout the silo. Unpalatable patches, resulting from soil contamination and poor fermentation, are most unwelcome as they tend to be rejected and fall to waste. Uneven removal of silage can also make high silo faces unstable, particularly when there is a narrow, unsupported column of material in a nearly empty silo. There is a risk of this with cattle when the silage face is higher than 2 m, or 1.2 m with ewes. The top layers are then beyond the reach of the stock, and must be thrown down to the bottom of the face for 'easy-feeding'.

Mechanised Feeding

Much attention is now being given to the mechanical extraction and feeding of silage from clamp and bunker silos. This offers a number of potentially useful features:

- there is greater flexibility in the siting of silos in relation to the animal housing because the silage can be transported.

- silage can be stored at a greater settled height than is possible with self-feeding, with unloaders able to deal with heights of 3 m or more.

- silage can be weighed and mixed with other feeds before it is fed.

- each silo can be used to feed several different groups of stock; alternatively two or more lots of silage can be mixed before feeding.

- unroofed clamps or bunkers with simple walls, which are cheap but not well suited to self-feeding, can be used.

Mechanised feeding of silage is being applied to all classes of dairy and beef cattle, and also to in-wintered sheep, and a wide range of equipment is now available for removing silage from the silo and delivering it to stock. Selection of the particular system must take account of a number of factors. Thus the size of the equipment depends on the number of animals, the quantity of silage to be fed to each animal, and the distance from the silo to the stock; the dimensions of the machine, and whether it is trailed, mounted or semi-mounted, will depend on building layout and access, width of feeding passages and height of feed troughs. The reach of the unloading equipment must be related to the height of the silage face. Finally, a forage box or mixer wagon is indicated if the silage is to be mixed with other feeds before it is fed, or if the amount fed is to be metered. The following paragraphs note features of the different systems available.

Silo Unloaders

Foreloaders, which are simple and versatile, are often used to unload from the silage face. However, unless carefully used they can open up the remaining silage, letting in air which can cause aerobic deterioration and moulding. Hydraulic grab attachments, which remove silage without disturbing the silage face, are now available for most loaders. They can also move silage from the silo to feed troughs with minimum spillage. Specialised handling vehicles, handling up to 20 tonnes of silage an hour (double the work-rate of conventional tractor foreloaders), are now used on many farms. While foreloaders can be used to transport silage over short distances they are more commonly used to fill some form of forage box or feeder.

Block Cutters (Plate 9.2)

These machines, which cut blocks of silage from the face of the silo with minimum disturbance, are mounted either on the rear of a tractor to reach a height of about 2.5 m, or on a foreloader, with a reach of up to 4 m. They leave the silage face compact and so minimise aerobic wastage. The blocks cut range in size from 0.85 to 2.5 m^3 (300 to 1,000 kg), and can be transported direct to the feed barrier where only occasional hand-forking may be needed as the block is eaten. The silage in the block remains fairly compact and in winter can be fed over several days with little deterioration. Some machines are also equipped to distribute blocks into the feed trough but are rather complex and heavy, and require a large tractor to operate them. Consequently machines have been developed to distribute pre-cut blocks as a separate operation. Some forage boxes are strong enough to handle blocks of short-chopped silage although this will add considerable stress to the mechanism if treated as a regular operation.

Self-Loading Feeders

These machines, mounted on the rear of a tractor, extract silage from the clamp and transport it to the feeding point. Some machines remove the silage by means of rotating cutters on the rear of the feeder, which is backed up against the silage face. Silage is cut as the unit is raised and

Plate 9.2   *Block cutters can handle all types of silage; they leave a compact and undisturbed silage face.*

lowered and thrown into the transport hopper from which it is then distributed by a cross-conveyor for feeding. On other machines the hopper is filled by a hydraulically operated combing action (Plate 9.3). All these machines will handle most types of silage, although they work best with short-chopped material. Most of them are tractor-mounted, typically with a 2 m³ transport hopper, hòlding 500 to 1,000 kg of loose

Plate 9.3   Feeders, filled by a hydraulically-operated combing action, cause little disturbance to the silage face (Photograph G. E. Amos).

silage, which is filled in three to five minutes. Although this capacity may limit its use for larger herds, the equipment is very manoeuvrable, and well adapted to smaller herds and to use in buildings with difficult access. Larger trailed machines, with up to double this capacity, are also available. All these machines have the particular advantage that they cause very little disturbance to the silage face. Block cutters and feeders are well-suited to dealing with unwalled clamps of short-chopped silage, which are now often made by contractors, and may then be the only machinery on the farm committed to silage.

Forage Boxes

For larger herds feeding equipment with greater capacity may be needed, and forage boxes, ranging from 3 to 10 m³ and holding up to 5 tonnes of silage, are commonly used. These are filled at the silo with a foreloader or grab; a conveyor on the floor of the box moves the silage forward to a set of rotating beaters which tease out the material on to a cross-conveyor for delivery to the feeding area. On some machines delivery can be to either side, which may be of advantage when access to

Plate 9.4 Forage boxes, generally filled by a tractor foreloader, and available in a wide range of capacities, are well suited to feeding silage to larger herds.

the building is restricted. Forage boxes can deal with most types of silage, including silage cut in blocks, but work best with fairly short-chopped material (Plate 9.4). Beater mechanisms consisting of large-diameter drums, each fitted with an arrangement of short teeth, are less susceptible to wrapping of long material than beaters with an open formation and longer tines. Several ingredients can be fed at one time with a forage box, but only limited mixing occurs. To achieve even moderate mixing the different ingredients should be loaded in layers so that they are mixed by the beaters as the feed is distributed; this is really only practicable with two or three feed components. One make of forage box can be fitted with a hopper on the front of the machine which delivers concentrates and other free-flowing ingredients on to the silage on the cross-conveyor.

Some types of manure spreader can be modified to feed silage loaded by a foreloader, by enclosing the beater mechanism and fitting a cross-conveyor or chute attachment to transfer the silage to the feeding area.

Mixer Feeder Wagons (Plate 9.5)

In cases where ingredients with very different physical characteristics must be mixed before feeding, special-purpose mixer wagons are preferred. Several different mixing systems are used. Many wagons are fitted with three or more augers, positioned horizontally along the wagon and moving in opposing directions so as to mix the load (Plate 9.6). The augers should be rotated slowly as the wagon is filled to avoid the heavy starting load of a full wagon. The mixed ration is discharged from a hatch in the base of the wagon and thence by a short retractable conveyor to the feed troughs.

Mixer wagons will deal with up to 10 per cent of the load in the form of long hay or straw, but mixing is less efficient than with shorter material. These machines will mix most ingredients to the point at which animals find it difficult to select individual components; however, animals can pick out large particles, such as whole roots or large cubes, unless these are chopped before being put into the mixer.

Mixer wagons range in size from 4 to 12 m³. The density of the mixed ration is typically 300 kg/m³, though there is considerable variation with different ration components. All types of forage boxes and mixer wagons can be fitted with integral equipment to weigh in the individual feed ingredients and to weigh out the mixed feed.

Plate 9.5   Mixer-feeder wagons will thoroughly mix and feed forages and other feeds.

Plate 9.6   In some mixer wagons, heavy-duty horizontal augers rotate in opposite directions to give good mixing of the feed ingredients.

Feed Troughs

The dimensions and design of arrangements for feeding silage need special care so that intake is not restricted while at the same time feed is not wasted. It is currently recommended that when forage/concentrate mixtures are being fed 0.7 m of trough length should be allocated to mature cattle. This can be reduced to 0.45 m for cattle six months old, increasing to 0.57 m for stock eighteen months old. Competition between animals is much less severe when the complete ration is fed *ad lib* and trough length can then be reduced to 0.15 m. Trough width should be related to the reach of the animals which is up to 0.8 m for a mature dairy cow eating from the base of a manger 300 mm above the feeding area. The base of the trough should therefore be raised 100–300 mm, with the front 0.5 m above this. The height of any wall constructed to retain feed should of course accommodate the chute of the feeder wagon which is invariably higher than 0.6 m though it is wise to check with the wagon concerned if larger dimensions are required. Feed troughs should have smooth internal surfaces free from obstructions that can harbour moulding feed and without uprights to interfere with discharge from the feeder.

Feeding Big-Bale Silage

When first introduced, bales of silage were fed by a variety of simple though imprecise techniques, each adapted to a particular building arrangement and feeding requirement. These included feeding whole bales from racks and unrolling bales along a feed fence, although the silage could be unevenly distributed. In some cases whole bales were placed in the feed passage and moved up to the stock as they were consumed. Stock housed in buildings without tractor access have been fed by hand-forking from bales sliced open with a 'hay knife', a method best suited to feeding small quantities of silage to housed sheep.

Feeding big bales of silage to sheep requires special consideration to avoid selective feeding and excessive waste without restricting intake. A number of 'special purpose' feeders are now marketed to meet these requirements.

As the technique has become established machines have been developed to emulate the mechanised feeding achieved with clamp silage and forage box systems. However, the long unchopped material in baled silage can present problems particularly for those machines designed to chop and lacerate the silage as it is distributed. These feeders tend to be complex, expensive and power-consuming. A more successful approach, more compatible with the system as a whole, has been to use machines which 'unroll' bales without chopping as they are cradled by a simple chain and slat arrangement (Plate 9.7).

Mechanised Feeding from Tower Silos

Some mechanised feeding systems are based on tower silos, with the silo unloader feeding into a forage box or mixer wagon, or alternatively on to a conveyor system. The latter have several advantages:

- feeding can effectively be fully automated, although precautions must be taken to ensure the safety of staff and livestock.

- there is no regular requirement for tractors and drivers.

- building space can often be more effectively used, because there is no requirement for wide feeding passages and turning spaces, and headroom is less critical.

Despite these advantages relatively few conveyor systems, either

Plate 9.7 Roll bales of hay or silage can be simply 'unrolled' for feeding by a system of chain and slats as the bale is carried in the tractor-mounted feeder.

from tower or bunker silos, have been installed. There are several reasons: conveyors are less versatile than forage boxes, because the materials handled must flow easily; a conveyor can be loaded from a bunker silo usually via a forage box, but this involves the use of tractors and it is difficult to get a uniform flow of silage; and while other feed ingredients can be added from a hopper on to the stream of silage, little mixing occurs. In the early stages of development there were also considerable problems with poor reliability of tower silo unloaders and conveyors. The latest equipment is much more reliable; most mechanical problems can be rectified in the farm workshop, while electrical breakdown in the unloading and feeding equipment and automatic controls can generally be diagnosed and rectified by a local electrical contractor. However, it is essential to have some alternative system available for feeding the stock in case a breakdown lasts for more than a few hours.

The pneumatic blowers used for filling and emptying some tower silos can also be used to convey the silage directly to the feed fence or troughs. This is done by linking the blower to a flexible telescopic pipe of 450 to 600 mm diameter, terminating in a cyclone from which the silage drops to the feed trough. Such a conveying system can deal with silage up to 100 mm long, and at up to 3 tonnes per hour, depending on the power available and the conveying distance.

Mechanical conveyors are more commonly used. Chain and slat conveyors are simple in design and can deal with a wide range of materials, but are susceptible to wear from abrasive materials. Long feeds can also wrap round the paddles. Thus belt conveyors are generally preferred, although there is still a risk of long materials wrapping on the rollers. They have a much lower power requirement than pneumatic systems, at about 0.5 kW per 10 m length of belt. The feed on the belt is transferred into the feed trough below by an angled plough or brush. Some systems also incorporate a continuous belt weigher which can weigh the amount of silage delivered to each group of stock.

A wide range of feeding systems for both hay and silage is now available; many of them achieve a high degree of mechanisation, allow other feed components to be mixed with the forage, and can be equipped with automatic weighing. Possibly the main emphasis is now being given to improving the *reliability* of equipment which must be able to supply feed to livestock daily over long periods. The importance of regular and careful maintenance just cannot be overstressed.

Chapter 10

FEEDING CONSERVED FORAGES

WE NOTED in Chapter 1 that hay and silage were in the past looked on as 'maintenance' feeds, and that the expectation that their feeding value would be low was at least partly responsible for the wasteful methods of conservation that were so often adopted. We believe that low-quality feeds, made with high losses, have little part to play in future feeding systems. Thus in the following sections we consider mainly the feeding of the better-quality conserved forages described in earlier chapters— hay and silage made from crops cut at a relatively immature stage of growth and conserved by methods which minimise losses and which give feeds of good intake and digestibility characteristics.

This chapter examines the potential of such feeds to give good animal production without the need for high levels of supplementary feeding, while the last chapter identifies some of the management and economic factors that must be considered in selecting the best conservation strategy.

Conserved Forages in Dairy Cow Feeding

Figure 1.1 showed that, until about 1972, average milk yields in the UK had increased at about 2 per cent a year, but that from 1972 onwards they increased much more rapidly. This remarkable improvement in production was clearly stimulated by entry into the EEC, which encouraged the faster adoption of improved animal breeding, housing and health control. It also encouraged better feeding, in which more efficient forage conservation played an important role.

One of the most significant advances in feeding was the development of the concept of 'lead-feeding' at the National Institute for Research in Dairying (NIRD) at Shinfield. Traditionally the amount of concentrates fed to dairy cows had been based on the cow's milk yield on the previous day; in contrast in lead-feeding the amount of concentrates fed in early lactation is steadily increased to a level beyond which there is no further increase in daily yield (hence the alternative name of 'challenge feeding'). The aim is to stimulate as high a peak yield as possible, for workers at NIRD had shown that total lactation yield is likely to be about two hundred times the peak yield.

In practice appreciably more concentrates are fed during early lactation in lead-feeding than in conventional systems of rationing. As a consequence cows are likely to eat less forage at that time, because of the substitution effect of concentrate feeding (page 40). If it is assumed that each 0.4 kg of concentrates fed produces a litre of milk, conventional calculation then indicates that forage is contributing only maintenance, or even less, to the total ration. This would not matter, except that on many dairy farms this was taken to represent the *potential* of the forage—and concentrates continued to be fed at close to 0.4 kg per litre throughout the rest of the lactation. Milk yield increased sharply, but at the cost of high concentrate use.

This problem was most effectively tackled in the Brinkmanship system of feeding, developed by Ken Slater, then at the Cheshire Farm Institute. This employs lead-feeding during early lactation to stimulate a high peak milk yield; but from about the fifteenth week of lactation onwards the amount of concentrates fed each day is cut down more rapidly than in conventional feeding, and the animals begin to eat more forage. Provided the forage available is of high D-value and intake potential it can then make up an increasing proportion of the total nutrient requirement; from the twentieth week onwards well-preserved silage can readily contribute 'maintenance' plus 15 litres daily to the total ration (this division into 'maintenance' and 'production' has of course no nutritional significance, but it remains a useful way of indicating the feeding potential of a forage). Thus although high levels of concentrates are fed in early lactation, by exploiting fully the potential of the forage part of the ration later in lactation, the *overall* amount of concentrates fed per litre will be significantly lower than with conventional rationing. Many individual lactations above 6,000 litres, but with less than 0.3 kg of concentrates fed per litre, have been recorded.

In systems based on lead-feeding, including the Brinkmanship system, the cows in the herd have to be individually recorded and fed. As average herd size, and the number of cows milked per man, increased during the 1970s, the necessity for individual recording and feeding began to be questioned, and new systems of allocating concentrates were sought. Many of these were based on the new concept of 'flat-rate' feeding; unfortunately this has led to some confusion, because this term has been used to describe several rather different feeding systems. These include: (a) systems in which the same amount of concentrates is fed daily throughout most of the lactation to all the cows in a herd, irrespective of individual level of milk yield; (b) systems in which each cow is fed daily an amount of concentrates based on her milk yield fourteen days after calving, which is taken as a measure of her potential milk yield; (c) systems similar to (b), with concentrate allocation based on fourteen-day yield, but with the level of concentrate feeding reduced at intervals during the lactation (sometimes called 'stepped feeding').

In our view these distinctions have not always been made sufficiently clear. Equally importantly, it has not always been emphasised that the cows in flat-rate feeding systems must have *ad lib* access to high-quality grazed pasture or conserved forage at all times. If that condition is satisfied, then option (c) seems likely to be the most effective. For, under system (a), if the level of concentrate feeding used is adequate for the high-yielding cows in a herd then it should be possible to feed less concentrates and more forage to cows with a lower yield potential; similarly under (b) there must be scope for feeding less concentrates later in the lactation, as in the Brinkmanship system, so as to make maximum use of forage as a cheaper source of ME than concentrates, yet without sacrificing total lactation yield.

Both experimental work and practical experience now confirm that the way a given amount of concentrates is allocated during the early part of lactation has little effect on total milk yield, as long as high D-value hay or silage is freely available. For example, work at the West of Scotland College has shown very similar yields, at 26.5 litres a day, from two groups of cows, one fed 9 kg of concentrates daily over 20 weeks and the other 11 kg daily, reducing in steps to 7 kg.

Much work has also shown the importance of the level of forage digestibility. Thus in one trial in Scotland cows fed on silage of 58.5 D-value gave 22 litres of milk daily, compared with 24 litres from another group fed the same amount of concentrates plus silage of 64.8 D-value—in line with other experimental results showing that milk yield increases by about 0.3 litres for each one unit increase in digestibility of the silage fed. Overall, for a given milk yield less concentrates are needed if more digestible silage is fed (Table 10.1).

Table 10.1 Target intakes for silage and concentrates by a 6,000-litre autumn-calving cow during 150 days of winter feeding, at different levels of silage digestibility

D-value of silage	61	64	67
Intake (*tonnes dry matter*):			
silage	1.7	1.9	2.0
concentrates	1.3	1.1	0.8

(Data: AGRI and ICI plc)

Some economic aspects in choosing the optimum combination of the amount of concentrates and the amount and quality of conserved forage fed are considered in the next chapter; but for most dairy cows the pattern in which the concentrates are fed appears less critical than was earlier thought—though there is little evidence that flat-rate feeding can support very high levels of milk production, such as the 9,729 litres, at 0.33 kg per litre, of one herd fed on the Brinkmanship system. But in our view there is likely to be advantage in feeding more concentrates to the potentially higher-yielding cows in a herd. ADAS suggest that milk yield on the fourteenth day of lactation can give a good indication of milk potential. All cows are fed the same level of concentrates, increasing up to fourteen days after calving. From then on each cow is fed concentrates at the daily 'flat rate' indicated in Table 10.2, based on her milk yield on day 14, and on the digestibility of the silage being offered. Feeding at this level is continued for about a hundred days, after which the amount of concentrates fed can be steadily reduced to turnout, as in the Brinkmanship system. It must be noted, of course, that lower-yielding cows will eat more silage, and that this must be set against any saving in the cost of concentrates fed.

Table 10.2   Suggested level of 'flat-rate' concentrate feeding of autumn-calving cows, based on milk yield at 14 days and D-value of silage fed

	Milk yield at 14 days (litres)			
	20	*22.5*	*25*	*27.5*
	(Estimated 305-day milk yield, litres)			
Silage	*5,000*	*5,600*	*6,250*	*6,875*
D-value	*Flat-rate concentrate feeding (kg per day)*			
61	8.6	9.7	10.7	11.8
65	7.0	7.9	8.7	9.6
69	5.0	5.6	6.2	6.9

(Data: ADAS)

For all cows the silage must also be fed in a way that does not restrict the amount eaten. This is one reason why there has been some shift away from feeding at the silage face, which can sometimes restrict intake, to systems in which the silage is cut and fed from troughs. Long silage can also reduce intake, though there seems no advantage in chopping to less than 50 mm. The silage must also be fresh; low-pH, moist silage can remain fresh for up to twenty-four hours, but silage above 25 per cent DM, and in particular maize silage, becomes stale fairly quickly and should preferably be cut and fed twice daily.

Silage is now the main form of conserved forage fed on most dairy farms. But very similar principles also apply when hay is fed; in particular the hay must be of adequate D-value, and must be offered in a way that does not restrict the amount eaten. There is also increasing interest in the use of alkali-treated straw (Chapter 8) in dairy cow feeding. Even when the straw must be bought in, treated straw may provide ME at similar cost to that of hay or silage made on the farm; on many mixed dairy/cereal farms, where straw is a surplus product, it can provide ME more cheaply than hay or silage. By using straw in this way less land is needed for forage production, and more land can be used for growing cereals—as long as there is no restriction on the amount of cereals that can be grown. Cereal straw can also be treated at any time of year, so reducing the demand for labour and machinery during the 'peak' period of forage conservation in May and June. Treated straw can be a useful supplement to silage, as its residual alkali content helps to maintain the intake of acid silages, in the same way as sodium bicarbonate, noted on page 41; it can also be a useful 'buffer' feed when there is a shortage of summer grazing, particularly for late lactation and dry cows.

Despite these many advantages, treated straw can only replace part of the forage in dairy cow rations, because it is only a medium-digestibility feed (say 55–60 D-value, with an ME up to 8.5), and this also limits the amount that animals can eat. But where an adequate supply of treated straw is available the aim must be to feed as much as possible (because it can provide ME at low cost) given that the stock get enough total nutrients to achieve their full production potential. Based on considerable practical experience with alkali-treated straw Gordon Newman has suggested that cows in early lactation, giving more than 30 kg of milk, can be fed up to 4 kg of treated straw a day, increasing to 6 kg in late lactation and to as much as 8 kg with dry cows. At an average daily intake of, say, 5 kg, this could reduce the amount of silage needed for an autumn-calving cow by as much as 2 tonnes.

Feeding treated straw does of course involve an additional operation in the daily feeding routine; some dairy cows may also not eat their full allocation. For these reasons this feed fits particularly well into systems of mixed ration feeding, using a forage box or mixer wagon (page 158) to mix the straw with hay, silage, cereals, and in some cases concentrates, before it is fed. This system gives full opportunity for high feed intake. At the same time it allows feed mixtures of different energy concentrations to be given to different yield groups, and for the energy concentration of the mixture to be reduced as the lactation progresses. Thus at Bridget's EHF the diet for high-yielding cows contains more concentrates and less forage than that fed to lower-yielders, while the ME of the mix fed to high-yielders is reduced from 11.9 MJ per kg DM down to 10.2 in late lactation, by increasing the proportion of forage to concentrates in the mix. Feed mixers also allow other ration components to be included, such as manioc and maize gluten, which may represent good value but are difficult to include in normal feeding routines.

The previous paragraphs have emphasised the importance of ensuring high intake, in particular of silage. It is thus perhaps surprising that, though wilting has been shown to increase the amount of silage stock will eat (page 38), experiments have revealed no significant advantage, in terms of milk production, when wilted and unwilted silages have been compared. This may be because both digestibility, and the efficiency with which the digested nutrients are used for milk production, are lower in wilted silage. In contrast many experiments have shown a consistent increase in level of milk production when silages treated with acid or acid/formalin have been compared with untreated silages (Figure 10.1). The conclusion is in line with that in Chapter 2, that wherever possible a limited amount of wilting should be carried out so as to improve the chance of making good silage and to reduce effluent from the silo; but if wilting means delaying the date of cutting then the choice must be to ensile without wilting, using an additive. Certainly there seems no advantage in wilting to above 25 per cent DM in making silage for dairy cows.

Even when hay or silage of high D-value are fed, considerable amounts of concentrates are needed to ensure high milk yields in early lactation. This can pose problems, in particular in the risk of digestive upset if all the concentrates are fed at milking times. In many herds effort is made to avoid this by feeding part of the cereal and concentrate allocation outside the parlour, from troughs adjacent to the forage feeding area. If different yield groups feed together there will also be advantage in giving the higher-yielding cows access to additional supplementary feed by using electronic feed dispensers. Mixing cereals and concentrates with the forage part of the ration before feeding is also an effective way of reducing the risk of digestive troubles. This might in theory also appear to provide the cow with a more uniform supply of digested nutrients through the day, but there is little evidence that mixing the forage and concentrates gives more efficient milk production. However, it does seem to minimise the reduction in milk fat percentage when fairly high levels of concentrates (more than 10 kg a day) are being fed. Concentrates are of course not a standard commodity and, as was discussed in Chapter 3, different forms of concentrate affect hay and silage intakes to different extents. Concentrates containing a high proportion of cereals give the greatest reduction in forage intake, but this effect is reduced if part of the cereal is replaced by a feed, such as

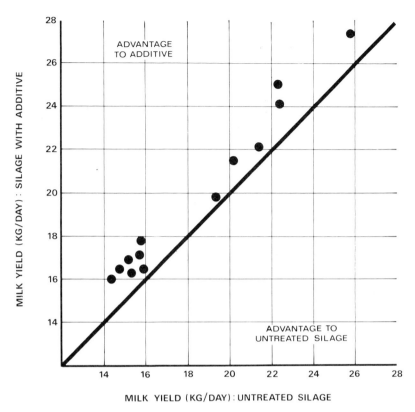

Figure 10.1   Milk production from silages made with and without additive (Data: Wilkinson)

fish-meal, with a high content of rumen-undegradeable protein. Thus work in Scotland and N. Ireland has shown that more silage is eaten, and more milk is produced, when the protein content of the concentrate is increased by replacing part of the barley by fish-meal—though at least part of this effect is likely to result from the higher intake of protein *per se*. A 'balancer' cube, developed at the Hannah Dairy Research Institute, containing 65 per cent of groundnut meal, and fed at 0.15 kg per litre, has given the same milk yield as 0.4 kg of barley, when fed as a supplement to high D-value silage plus brewer's grains. However, the cows fed the 'balancer' have eaten 1½ to 2 tonnes more silage during the winter, and the 'balancer' also costs more per kg than barley. Such factors must be considered in deciding the economic benefit of measures taken to increase the proportion of silage that can be fed.

There was also interest during the early 1970s in feeding dried grass cubes as a supplement to silage. Because of the high content of un-degradeable protein, resulting from the heat treatment during drying, and the small particle size in the cubes, there was very little reduction in silage intake, and good levels of milk production were recorded. How-ever, the present low output of dried grass means that few dairy farmers are now able to use this potentially valuable feed.

The total energy intake of the dairy cow can also be increased, without reducing the amount of hay or silage that is eaten, by feeding supplementary fat, in the form of prills. These are mainly digested in the hind-tract, and so do not interfere with rumen digestion. Up to 1.5 kg of fat can be fed daily, either mixed with the hay or silage fed in troughs, or very effectively through a forage box or mixer wagon.

Conserved Forages for Beef Cattle

Systems of beef production have changed greatly in the last forty years. The expansion of the United Kingdom dairy industry, and in particular the shift to Friesian cattle, with their good beef potential, has resulted in some two-thirds of home beef supplies now coming from the dairy herd, rather than from specialist beef herds. Consumer demand has increas-ingly been for leaner meat, and this has been encouraged by medical advice to reduce the consumption of animal fat. For many years this trend was discouraged by official grading standards (on which price support for beef production has been based) which demanded a high level of 'finish' in beef carcasses, but standards are now more in line with consumer demand. Undoubtedly, though, this discouraged the wider use of crosses using later-maturing breeds—Charolais, Limousin and South Devon, for example—which give faster growth but are more difficult to finish than earlier-maturing crosses.

To produce the lean cattle, in the 250 to 280 kg carcass range, which much of the market now seeks, animals must grow fast; in particular they must avoid the winter 'store' period of most earlier feeding systems, during which little liveweight gain was made, and for which hay, silage and straw of only 'maintenance' quality were adequate feeds.

Because of problems in making conserved forages of high nutritive value, the first response was to feed increasing amounts of cereals during the winter, as in the 'barley-beef' system, developed at the Rowett Research Institute in the early 1960s, in which only cereals were fed. However, this did not become the dominant system of beef produc-tion in the United Kingdom, as was predicted at the time. Barley-beef has continued to serve a premium market; but its main effect was to encourage the development of alternative feeding systems which give faster and more consistent gains than traditional systems, but which use less cereals—a high-cost feed resource within the EEC—and which produce heavier carcasses than barley-beef feeding, so spreading the initial cost of the beef calf over more output.

Most beef cattle are now slaughtered at between 14 and 24 months of age, in contrast to the 24 to 36 months and older of previous systems. This means that cattle must make good daily rates of gain throughout their life, both winter and summer; to do this conserved forages have had to play a more positive role than before.

Yet improved methods of forage conservation have not been adopted as widely or as fast on beef as on dairy farms—at least in part because beef enterprises are in general smaller than dairy enterprises, and have not seemed to justify the same level of investment in equipment and management skills. But research has continued to open up new technical options in beef feeding. Coupled with the considerable practical experience now available, this has led to a bigger contribution by conserved forages in beef production.

The Importance of Forage Quality
If a beef animal is to make good liveweight gain it must have a high intake of nutrients—it must be able to eat large amounts of food of high digestibility. The benefits of high digestibility were shown in Table 3.3, with overall daily gain by cattle increasing by about 25 g for each unit increase in the D-value of the forage fed. The response to higher digestibility in silage is similar, with a mean response from many experiments of about 34 g extra daily gain for each unit increase in D-value. But this will only happen if the silage is well-fermented, for the intake of badly preserved silage will be low. Thus work at the Irish Agricultural Research Institute showed that the intake of a badly fermented silage (pH 4.8, high ammonia content) was only 6.3 kg of DM, and daily gain 0.47 kg, compared with 8.5 kg of DM, and 0.89 kg gain, when silage made from the same crop, but with formic acid used, was fed (pH 4.2, and low ammonia). In twenty-six similar comparisons of silages made with and without formic acid, the acid has given an average increase in daily gain of 0.18 kg (Figure 10.2).

Silage fermentation and intake can also be improved by wilting, and experiments have shown daily gains some 0.12 kg higher when wilted silages have been fed. Comparisons between wilted silage and silage made with formic acid at the Agricultural Institute have also shown very similar gains. Additives then can offer an alternative to field wilting in making silage for beef cattle, though for reasons already noted some degree of wilting should be practised wherever possible.

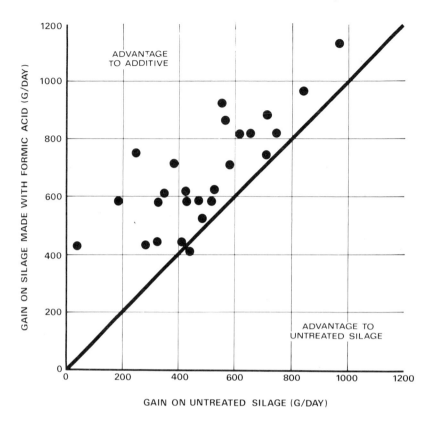

Figure 10.2   Daily gains by beef cattle fed silages made with and without additive
(Data: Wilkinson)

The Feeding of Cereal Supplements
Cattle fed on silage as the sole feed will seldom gain more than about
0.8 kg a day, and often considerably less; this is a marked improvement
on much earlier feeding, but is not always adequate for modern beef
production, in which gains nearer to 1 kg a day are generally sought.
Thus most beef feeders now feed supplements with their conserved
forages, particularly during the later stages of feeding. Most supplements
are based on cereals, which reduce the amount forage animals are able
to eat, so that the response to supplementary feeding is often less than
would be predicted, particularly with forages of high D-value. Thus in
one experiment 1 kg of a cereal supplement gave close to the 0.25 kg
expected extra daily gain when fed with silage of 58 per cent D-value,
but only an extra 0.1 kg with a higher quality silage of 65 D-value. The
intake of silage made from wilted crops, and from crops treated with

additives, is also reduced more by cereal supplements than that of the corresponding untreated silage. This means that some of the advantages of higher intake and higher gains when treated silages are fed tends to be lost when they are supplemented with cereals. Despite this there must be advantage in aiming for the best possible fermentation and intake potential, particularly when only small amounts of cereals are to be fed.

Protein Supplements for Beef Cattle
Not all supplementary feeds cause as much reduction in the amount of forage eaten as do straight cereals; in particular high-protein supplements reduce forage intake less than cereals (Chapter 3), and so should allow forage to contribute more to the total ration. However, only younger beef cattle are likely to benefit *nutritionally* from protein supplements, for the protein requirements of older cattle can generally be supplied from forage alone. In contrast, silage fed to younger cattle may be deficient in protein because most of the crude protein it contains is highly soluble and so relatively inefficiently used in the rumen. As a result there have been good responses to protein supplements fed with silage to beef cattle up to 150 kg liveweight, with daily gains doubled, using supplements of soyabean meal and fish-meal, whose protein content is only partly digested in the rumen. Hay or dried grass, in which the protein is less soluble than in silage, can also provide useful supplements.

Older cattle, though they have lower protein requirements, also respond better to protein than to cereal supplements because the former allow more forage to be eaten. For example, in farm trials in Northumberland 1 kg of a supplement containing either soyabean meal or fish-meal gave higher daily gains than the same amount of mineralised barley (Table 10.3).

Table 10.3 Daily gains during winter (184 days) by yearling cattle fed silage ad lib plus 1 kg of cereal or protein supplement

Supplement	Initial liveweight (kg)	Final liveweight (kg)	Daily gain (kg)
Barley	210	288	0.4
40% barley/60% soyabean	218	394	0.9
75% barley/25% fish-meal	220	400	0.95

(Data: ADAS Northern Region)

The possible benefit from restriction of the rate of protein digestion in the rumen when a formaldehyde-based silage additive is used was noted on p. 23. Possibly because of the relatively low protein requirement of most beef cattle experimental work has generally shown little advantage when formaldehyde/formic acid silages have been compared with silages made with formic acid alone. However, in feeding trials at the East of Scotland College of Agriculture, Charolais cattle of high growth potential gave daily gains of 0.98 kg when fed the mixed-additive silage, compared with 0.83 kg on formic-acid silage. Interestingly the 'formic-acid' cattle were also fatter at slaughter. This is in line with practical experience that cattle tend to put more energy into fat deposition when their main feed is silage, compared with hay or dried grass made from the same crop. This could be because silage, containing highly soluble crude protein, may be marginally deficient as a source of protein, so that more of its digested energy content is available for fat synthesis. Formaldehyde-treated silage, in which the protein is less soluble, may encourage protein rather than fat synthesis.

Conserved Forages for Sheep

Conserved forages have traditionally played a smaller part in sheep feeding than in beef or dairy feeding because most sheep have grazed outdoors throughout the year. Supplementary feeding, mainly with low-quality hay or straw plus limited concentrates, has been confined to ewes during pregnancy and lactation, and to other sheep during periods of shortage of grazing, as when the ground is snow-covered.

However, with the current trend to more intensive management of sheep more conserved forage is now being fed. Thus higher stocking rates on both upland and lowland pastures are leading to shortage of forage for grazing during the winter months, and this is being accentuated by higher lambing figures—and so higher nutritional demands—during the same period. Thus more supplementary feeding is needed. This has become more practicable because of the wide adoption of winter housing of sheep, which has simplified the operation and control of hand-feeding. Indoor housing also improves the management and supervision of sheep, and has improved both ease of lambing and lamb birth-weights. It does however mean that the whole winter feed requirements of the sheep must be supplied. Because hay and silage can provide ME at something under half the cost of cereals the aim must then be for conserved forage to make up as much as possible of this winter feed.

The provision of forage to be conserved for winter feeding does pose some problems. On lowland farms the ewes and lambs will be moved out from housing to pasture as soon as enough grazing is available and,

as long as lambing is not very early, the increase in daily feed requirements as the lambs grow will be in phase with the more rapid pasture
growth during April and May (Figure 4.1). At the same time a part of
the grassland area can be set aside to be cut for hay or silage for the
following winter; the regrowth on these areas will provide the 'clean'
grazing, free from parasitic worms, needed by the lambs after weaning.
However, with the present trend towards earlier lambing there is now
greater pressure on spring grazing, and more nitrogen is being applied
to sheep pastures to increase herbage growth for both grazing and
cutting. Despite this there is often insufficient grass to be cut for
conservation, and increasing importance is being placed on conserving
grass later in the season, when the demands for grazing by the ewes are
less. Such grass is often available in smaller 'lots' than in the spring; in
summer this can often be stored as hay, but on many sheep farms
big-bale silage has now been adopted as an effective way of conserving
fairly small and intermittent batches of grass. Regrowths from the cut
areas can provide the highly digestible grazing needed to flush the ewes
in autumn.

Sheep on hills and uplands pose greater problems. Previously it was
most often the winter carrying capacity that determined the number of
sheep that could be carried on upland farms. But with more winter
housing the amount of forage that can be cut for conservation can now
often be the limiting factor. In many cases this is not enough for the
number of sheep that can be grazed during the summer, and supplementary feeds must be brought in. ADAS trials indicate that it may then be
more economical to purchase straw rather than hay, for the daily cost of
feeding a ewe in winter on purchased hay plus 0.7 kg of concentrate is
likely to be more than on straw plus the 1.0 kg of concentrate needed to
provide the same amount of energy.

On most farms the aim must be to produce as much as possible of the
conserved forage that will be needed for the winter feeding. Traditionally
forage has been conserved as hay, reflecting the widely held view that
silage is not a suitable feed for sheep. It is now clear that this view was
more a reflection of the poor quality of the silage that was made on
sheep farms rather than of any basic deficiency in silage as a feed for
sheep (cattle on the same farm would in any case probably have had first
choice). With the adoption of improved silage-making methods excellent
results are now being obtained with silage fed to sheep.

This is fortunate; for most of the sheep in the United Kingdom are in
the north and west of the country, and many of them in the uplands,
areas which are much better suited to silage than to hay as the main
method of forage conservation. The practical success that has been
achieved from feeding well-fermented silage to sheep is thus of great

importance. There will of course be advantage if the cut crop can be wilted before it is ensiled; but wilting just cannot be relied on in the parts of the country where most sheep are congregated, and there is a very strong case for the use of silage additives in these areas, if necessary at an above average rate so as to guarantee the good fermentation needed to ensure high intake when the silage is fed to ewes. As on lowland sheep farms, much of this silage is likely to be made in relatively small batches, and there has been a remarkable expansion in the amount of big-bale silage, much of it made by contractors, in upland areas. The main disadvantage of big-bale silage for sheep is that the grass is not chopped before it is ensiled, for the amount of silage that ewes can eat may be limited if the grass is very long, and particularly if the bales are very dense. Thus every effort must be made to ensure that access to the silage is as easy as possible, yet without wastage.

Ewes are likely to remain at grazing for the first two or three months of pregnancy, and when first housed require feed of only medium quality. Hay or silage of quite low D-value—in the 50 to 55 range—should be adequate, possibly supplemented with a feed block containing non-protein nitrogen such as urea. Alkali-treated straw (Chapter 8) could be useful at this time, though some ewes may not eat this feed very readily. During the last two months of pregnancy forage quality is more critical, because the ewe's capacity for feed is restricted by the volume of the foetus just at the time that her nutrient requirements are increasing fast. Thus to keep the amount of concentrates fed at a reasonable level the hay or silage being fed must be of high D-value. Recent experience has also shown the great importance of avoiding soil contamination in silage to be fed to sheep, to prevent infection with listeriosis.

Hay of 77 D-value has been fed experimentally to provide the whole nutrient requirement of a ewe carrying twin lambs; but to make hay of this quality, even with barn-drying facilities, would in general be quite impractical. A more realistic target is hay or silage of around 65 D-value. Conserved forage of this quality, fed with 450 g of concentrates a day, should be fully adequate for a ewe with twins—provided, if silage is being fed, that fermentation was good. Conserved forages of lower digestibility will need higher supplementary feeding, which in turn will tend to reduce the amount of forage the ewe can eat. Thus as much as 0.75 kg of concentrates a day will be needed when hay of only 50 D-value is fed (Table 10.4). This is high-cost feeding; it can also lead to risk of digestive upset, and every effort must be made to conserve better forage than this.

Once the lambs are born forage intake will be less limiting than with the pregnant ewe, as a result of the effective increase in the ewe's

Table 10.4 The amounts of hay or silage of different digestibilities, and of concentrates, eaten by a ewe with twin lambs during the last 5 weeks of pregnancy. The calculated ME contents of the diets, in relation to the daily requirement of 16 MJ ME, are also shown

Conserved forage	Hay	Hay	Hay	Silage
D-value	77	64	50	68
Intake, g DM per day:				
conserved forage	1,630	1,115	560	1,190
concentrate	0	370	740	370
Estimated ME in diet				
(MJ per day)	19.8	16.0	13.6*	17.0

*This diet would be deficient in energy.
(Data: AGRI, Hurley)

capacity for food. Thus if several lots of hay or silage of different qualities are available for winter feeding the best lots should be allocated for feeding in late pregnancy. But to keep overall concentrate feeding below 500 g a day a forage quality above 60 D-value should be aimed for.

This chapter has examined the technical possibilities for meat and milk production from rations containing a high proportion of conserved forage, and has indicated the high levels of production that can be obtained. Less attention has been given to identifying the feeding systems that are likely to give the best financial returns under different farming situations; for a system that places the greatest reliance on conserved forage is not necessarily the one that makes the most profit. For example, higher winter gains such as those recorded from feeding a protein supplement to beef cattle (Table 10.3) would only be profitable in cattle sold for slaughter directly from winter feeding; if they had a further period of grazing before slaughter the unsupplemented group, which made lower winter gains, would almost certainly make most of this up through cheap 'compensatory growth' at pasture.

Thus the following chapter examines some of the management and economic factors which must be considered in determining the optimum conservation strategy for particular livestock enterprises, as well as the changes that may be needed in response to possible future changes in prices and market demand.

Chapter 11

FORAGE CONSERVATION IN FARMING SYSTEMS

CHAPTER 1 outlined some of the considerable changes that have taken place in forage conservation in the United Kingdom in the fifteen years since this book was first published. Overall there has been a steady increase in the amount of forage conserved each year; with little change over the same period in the numbers of livestock this means that, on average, more conserved forage is now being eaten by each animal. More significant has been the marked swing from hay to silage as the main method of conservation (Figure 1.2). This is continuing; a survey during the wet summer of 1985 showed an increase of 18 per cent in the area cut for silage, and a further reduction of 5 per cent in the area cut for hay. There is little doubt that the substitution of silage for hay as the main winter feed has played an important role in the sharp rise in milk production per cow in recent years. For, although considerably more concentrates have been fed over the same period, *there has been a steady decrease in the amount of concentrates fed for each litre of milk produced* (Figure 11.1), indicating an increasing contribution by forages to the total ration. Similarly the greater amounts of hay and silage now being fed to sheep has been stimulated by, and has certainly contributed to, the greater profitability of sheep husbandry, and silage-based systems of feeding beef cattle remain competitive with more intensive cereal-based systems.

Better conservation must have contributed to the major problem now facing United Kingdom and European agriculture, that of excess food production; equally, in our view it can now make an important contribution to dealing with that problem. As we noted in Chapter 1, the first really effective action taken to check surplus production in the EEC was the introduction of milk quotas in 1984. No longer were dairy farmers able to compensate for lower real prices for milk by producing and selling more; they now had to produce their quota as cheaply as possible. Many dairy farmers quickly responded by feeding more forage and less concentrates—and in many cases found that the resulting small fall in milk output, needed to respond to the quota, was compensated by a lower feed cost per litre of milk produced (Table 1.2).

This is in line with the often-repeated slogan that 'grazed and conserved forages are the cheapest forms of ruminant livestock feed'. But we must remember that this has been at least in part because the main alternative feeds, the cereal-based concentrates, have been expensive because of the high-price EEC cereal regime—a regime which, in its turn, has led to the present huge cereal surpluses in the Community. There is now general agreement that action must be taken to curb further surplus production; *but the way this is done could have a*

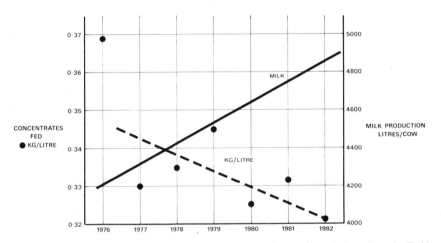

Figure 11.1   Average British milk yield per cow, and kg concentrates fed per litre of milk (the high figure in 1976 resulted from the severe drought)

considerable effect on future grass and forage feeding systems. Control of cereal output by some form of physical restriction on production, for example by quota or set-aside—even if linked with continuing price restraint—would probably mean that EEC cereal prices would continue at above 'world' levels, and so remain relatively uncompetitive with well-managed forages. In contrast, control of cereal output by really drastic price cuts would make cereals, and the concentrates made from them, more competitive—and could well reverse the present trend towards greater use of forages. For, as we noted in Chapter 1, forage-fed ruminants may, at present, be more profitable. But growing and utilising forages demand much skill in management, and if 'simpler' feeding systems, largely based on cereals, were also equally profitable, they would almost certainly be more widely adopted.

This is only one of the considerations that make plotting the future course of forage production in United Kingdom agriculture so difficult. Just as this final chapter was being written the price of oil fell by more than half. This *must* reduce the cost of many types of agricultural inputs, including agrochemicals (in particular N fertilisers), machinery, costs of operating tractors and implements, crop drying, plastic sheeting, and so on. What is less clear is the effect that lower oil prices are likely to have on the costs of growing and using forages compared with the costs of producing cereals. It does seem certain though that a reduction in the price of N fertilisers could lead to more being used, and could make more likely some form of restriction on N use, so as to check overall farm output. This would reduce cereal production; it would certainly reduce grassland production, which has become so dependent on N inputs; it might also encourage the greater use of legumes, which can 'fix' their own N, and of forage maize, which can get most of its N from cattle slurry.

It is because of this *uncertainty* about the future framework within which agriculture will have to work, even a few months ahead, that *Forage Conservation and Feeding* has dealt mainly with technical options, rather than seeking to give specific advice. Thus the discussion of the relationships between yield and digestibility in Chapters 3 and 4 indicates when a particular forage variety should be cut to get a particular level of D-value; it does not advise which level of digestibility to aim for. The discussion on the interactions between conserved forages and different types of supplementary feeds in Chapter 3 could help the individual farmer to make a more effective choice of feeds; it does not advise him which feeds should be given, or at what level. Similarly Chapter 8 has described how the D-value of straw can be improved by chemical treatment; whether processed straw can be fed profitably will depend largely on its cost, relative to the current prices of alternative feeds.

Despite these uncertainties, however, there are some management and economic factors that seem likely to be as relevant over the next fifteen years as they have been over the last fifteen. Thus the cost of harvesting and storing a tonne of forage DM is likely to be similar for forages differing widely in quality. As a result the cost of storing a unit of ME becomes greater as forage quality falls, so that there must be advantage in harvesting at higher rather than at lower levels of D-value. Similarly, whatever type of forage is being conserved, the 30 per cent or higher losses that were the norm in the 1960s are no longer tolerable— hence the attention given to reducing losses in different methods of conservation. Most ruminant animals will also continue to be fed on mixed rations, though with the proportions of forages and other feed components changing in response to their relative costs at a particular time—hence the importance to the livestock feeder of knowing how different feeds interact and complement each other.

The availability of *practical* systems of conserving forages of predictable digestibility, and with relatively low levels of loss, should greatly extend the options available to livestock farmers to respond to the future changes in political and economic conditions in which they will have to operate. In particular it could lead to some reassessment of the relative roles of grazing, and of cutting for conservation. Traditionally grazing has been considered the cheapest form of livestock feeding, and this has led to emphasis on growing early and late grass so as to 'extend the grazing season', growing lucerne to withstand summer drought, etc., with the aim of making the time stock are fed indoors on conserved forages as short as possible. It was difficult to challenge this thesis as long as most conserved forages were of low feeding value and had been made with high losses. But now that efficient methods of conserving high-quality feeds are widely used the relative roles of grazing and conservation can be more sensibly discussed. In particular more attention is being given to the effective integration of grazing and cutting, in systems of animal production that are less weather-dependent than many current livestock systems.

Probably the most important development has been in the concept of 'buffer feeding'—the planned feeding of hay or silage to grazing livestock. Hay has of course often been fed in the past, to overcome an emergency such as shortage of grazing during a period of summer drought. But in buffer feeding, as pioneered by David Leaver and his colleagues at the West of Scotland College, conserved forage is fed as an integral part of the grazing system, often right throughout the grazing season. Thus in spring a daily supplement of 2 kg of hay or 8 kg of silage allows cattle to be stocked at high grazing pressure, so that a high proportion of the grass grown can be eaten without any fall in daily milk production or rate of liveweight gain. This means that a bigger area of first-growth forage—the best feed of the year—can be left to grow on to high yield for cutting in the second half of May. For dairy cows a supplement of hay or silage in spring can also provide the extra 'long fibre' in the diet needed to keep up butter-fat levels. The low seasonal price for milk in spring has also persuaded some dairy farmers to continue 'indoor' feeding to the end of May, so that as much as possible of their first-growth forage can be conserved and fed to give maximum milk production in the following autumn and winter, when prices are higher. This is, of course, in direct contrast to the concept, exemplified by the use of T-sums (page 57), of applying extra N fertiliser to grow grass for early grazing, so as to shorten the period of indoor feeding.

However, experience in the late spring of 1986 has again shown the unreliability of 'early grass'; furthermore the response to a unit of N fertiliser is lower than later in the spring, and taking early grass in this way risks reducing subsequent forage production on the sward. Thus on most farms we believe that the preferred policy should be to delay turnout in the spring, with the aim of growing enough forage to be conserved and fed in the following winter, to ensure that early turnout will not be needed the following spring. Always provided, of course, that an efficient method of conserving forage with low losses is being used.

The main role for buffer feeding, though, is later in the summer, when supplementary feeding with hay or silage allows stock to continue grazing on swards which alone would not provide enough food. One way of doing this is to allow the animals to graze during the day and to pen them overnight, when the conserved forage, up to 50 kg of silage (10 kg of hay) a day, is fed. This represents about a tonne of silage for each twenty dairy cows. Big-bale silage is well suited for buffer feeding, because its use can avoid the wastage and deterioration that can occur if silage is removed daily in small lots from a larger silo during the summer. Big-bale silage may also have been made from small lots of surplus grass cut during the previous autumn; though not of top quality this may be fully adequate for feeding to late-lactation dairy cows in July and August.

The ultimate extension of buffer feeding is of course 'storage feeding', in which stock are fed throughout the year on conserved forages and graze, if at all, for only a short period of the year. The main advantage of this system, best suited to larger units, similar to the 'feed lots' of the United States, is that it allows complete-diet feeding (page 167) to be fully exploited, by combining forages with a range of other feed components in 'least-cost' diets. But it does require a high level of mechanisation and, because it does not benefit from the *undoubted* low cost of summer grazing, it must be operated very efficiently to compete with systems which include grazing.

Storage feeding has mainly been used with dairy cattle, but several beef units have adopted the system developed at Rosemaund EHF, in which cattle are fed from ten weeks of age until slaughter at twelve to fourteen months on silage plus up to 2 kg of concentrate daily. The key to success with this system is the production of high yields of silage of 70 D-value and about 25 per cent DM. Such indoor feeding is particularly suited for bulls, for which statutory requirements for fencing make grazing difficult. Friesian bulls have given high growth rates and high feed conversion efficiencies on this ration, and have proved the most profitable to feed. Storage feeding also has advantages over grazing for steers and heifers, with which it is often difficult to maintain satisfactory rate of gain at grass as a result of fluctuations in herbage growth and quality. Some commercial units have not achieved the same success as at Rosemaund, most probably because silage feeding value has been too low as a result of delayed cutting.

Within all forage conservation systems the debate about the relative importance of forage quality and forage quantity will continue. In practice of course there is no single answer; quality and quantity will have different importance for different enterprises and for different classes of stock, and the optimum balance will also depend on the cost of alternative feeds, in particular cereals. But with silage the main conservation method, and with the recognition that a level of wilting above 25 per cent DM is seldom necessary, the decision *when to cut* can now much more readily be based on the optimum yield/digestibility of the crop than on the weather forecast. Date of cutting is made even less weather-dependent by the availability of a range of *effective* silage additives, whose efficiency is certain to be further improved.

Conserving forage has always been used as a way of getting over the problem of the seasonal pattern of grass production posed in Figure 4.1, by allowing forage to be carried over from periods of surplus to periods of shortage. Traditionally this was a most unrewarding operation, because of the high losses in the process and the low feeding value of most of the forage stored, which meant that high levels of supplementary feeding were needed to get satisfactory animal performance. The transformation that has been brought about by the wide adoption of better conservation methods in United Kingdom farming is only just beginning to be recognised—that hay and silage can now be made with low losses and that the products can have a feeding value equal to good grazed pasture. Cutting for conservation also makes it possible to *grow and to use* larger amounts of forage per hectare than can effectively be used by grazing livestock; against this must be set the inevitably higher cost, in harvesting and storage, of conserved than of grazed forage. But modern conservation methods do allow an objective assessment, and then application into practice, of livestock feeding systems which seek the best combination of grazed forage, conserved forage, and supplementary feeds. By exploiting the facility of conserved forages to compensate for variations in forage production between seasons and between years, meat and milk production from grass should perhaps become more reliable operations than they have been in the past.

INDEX

Additives, 19, 135
  AIV acid, 17, 19
  ammonium propionate, 13, 19
  applicators, 20, 135
  care in handling, 135
  effect on feed value of silage, 38, 168, 171
  formaldehyde (formalin), 23, 136, 174
  formic acid, 20, 38, 168, 171
  inoculants, 24
  molasses, 19
  sodium acrylate, 25
  sulphuric acid, 23
  to hay, 13, 96
  to silage, 20, 38, 135
Agricultural Development and Advisory
    Service (ADAS), 22, 40, 111, 166, 173, 175

Bale, as a feed unit, 150
  barn drying, 100
  density, 86
  handling, 81
  large round, 94
  large square, 94
  loading, 93
  moisture content, 86, 96
  size and shape, 85
Balers, 84
  output, 81
Barley, effect on forage intake, 40, 172
Barn-dried hay, 97
  feeding, 151
Barn hay-drying, 11, 97
  effect of air humidity, 98
  large-bale tunnel, 102
  loose hay-drying, 102
  storage driers, 100
  use of heat, 99
Beef production, 170
  feeding hay, 170
  feeding silage, 170
  Rosemaund system, 182
Big-bale silage, 130, 160, 182

Bridget's EHF, 167
Brinkmanship, dairy cow feeding, 164
British Association of Greencrop Driers, 15
Buckrake, for loading silos, 122
Buffer feeding, 181
Bulls, feeding for beef, 182
Bunker silos, 119
  dimensions for feeding, 152
  drainage, 137
  methods of filling, 122
Butterfat, effect of fibre in feed, 43, 168, 181

Cattle, beef, 170
  dairy, 163
Cereals, EEC prices, 179
Cheshire Farm Institute, 164
Chopping, for silage, 107
  relation to feeding methods, 153
Clamp silage, 128
Clover, red, 34, 38, 49, 55
  white, 34, 49
  see also legumes
Common Market (EEC), effect of policy on
    feed costs, 179
Complete ration feeding, 40, 167
Conditioning, combined with mowing, 65, 76
  field, 69
Consolidation of silage, 16, 123
Crimping, 65
Crop, digestibility, 28, 47
  for grass drying, 62
  for hay, 54
  for silage, 56
  planned spread, 52
  regrowth, 52
  yield, 50
Crushing, 65
Cutting date, effect on crop digestibility, 28, 47
  effect on crop regrowth, 52
  effect on crop yield, 50
Cutting methods, see mowers

Dairy cow feeding, *see milk production*
Dehydration, 15
  effect on nutritive value, 36
  pre-wilting, 15
  see also grass-drying
Digestibility, 27
  calculation of ME, 32
  conserved forages, 31
  D-value, 28
  effect of field treatment, 32
  effect on feed intake, 33
  estimation, 28, 32
  in vitro, 28
  maturity of crop, 28, 47
  measurement, 28
  of forages, 28, 47, 165
  of legumes, 48
  of maize, 59
  of permanent pasture, 31, 52
  optimum, 53, 165, 180
  relationship with fibre content, 28
  relationship with intake, 33
  stage of crop maturity, 47
Dorset-wedge method of filling silo, 119
  sealing, 120
Dried grass, intake, 36
  interaction with silage, 169
  particle size, 36
Drier operation, advantages of scale, 62
  benefit of pre-wilting, 15
Drying, baled hay, 100
  chopped hay, 102
  in the swath, 30, 65, 76
  large bales, 101
  'load', 9
Dry-matter (DM) losses, 80
  in hay, 10
  in silage, 17, 137

East of Scotland College of Agriculture, 174
Effluent loss from silos, 18, 137
Ensilage, *see silage*
European Economic Community (EEC), 179
Experimental Husbandry Farms (EHF), 37, 41,
  167, 182

Fertiliser nitrogen, 56, 180
Feeding, methods, 150
Feed intake, 33
  difference between forage species, 34
  dried grass, 36
  effect of feed supplements, 40
  hay, 34
  silage, 37
Feed troughs, 159
Fibre, and forage digestibility, 28, 43
  and milk butterfat, 43, 181

Field beans, for drying, 62
Flail, harvester, 108
  mower, 69
Forage box, 156
Formalin, silage additive, 23, 168, 174
Formic acid, silage additive, 21, 168, 171
  effect on nutritive value, 38, 171
Fuel consumption of driers, 15

Grazing management, 46, 181
Grass drying, 15
  crops for drying, 61
  cost aspects, 15
  economy of scale, 62
  for arable break crops, 15
  length of drying season, 62
  prewilting, 15
Grassland Research Institute, Hurley (GRI,
  now AGRI), 29, 30, 36, 40, 177
Grazing, 29, 46, 174
  buffer feeding, 181
  integration with conservation, 46

Hannah Research Institute, 38, 40, 76, 169
Harvester, double chop, 109
  flail, 108
  maize, 117
  precision chop, 110
  self-loading forage wagon, 115
Hay, additives, 13, 96
  baling, 84
  barn drying, 11, 97
  conditioning, 97
  crops, 54
  digestibility, 12, 32
  feeding methods, 150
  for sheep, 175
  heating in store, 10, 100
  interaction with other feeds, 40, 167, 176
  moulding, 10
  propionic acid additive, 13, 96
  proportion of conserved forage, 2, 178
  storage moisture content, 8, 86, 96
  wastage in feeding, 150
Haymaking, 8, 84
  chopped hay drying, 103
  drying systems, 97
  effect on feed value, 12
  loss of DM in field, 10
High-temperature drying, *see grass drying*
Hillsborough Agricultural Institute, 138

ICI Ltd, 25
Inoculants, for silage, 24
Intake, 33
  dried grass, 36
  effect of feed supplements, 39

Italian ryegrass, 35
legumes, 34
silage, 37
Irish Agricultural Research Institute, 171

Kale, for silage, 61

Large bales, 94, 160, 176
Legumes, digestibility, 48
 for hay, 55
 for silage, 56
 intake, 34
 lucerne (alfalfa), 34, 49
 mineral content, 44
 nitrogen fixation, 57
 red clover, 34, 38, 55
 sainfoin, 34, 48
 white clover, 34, 48
Losses, during feeding, 150
 in haymaking, 10
 in silage effluent, 18, 137
Lucerne (alfalfa), *see legumes*

Maize, cultivation, 58
 D-value, 59
 for silage, 60, 134
 harvesting, 117
 non-protein nitrogen additives, 135
Maize Development Association, 58
Metabolisable energy (ME) of forages, 56
Milk Marketing Boards, 4
Milk production, 3, 163
 Brinkmanship system, 164
 flat-rate feeding, 164
 lead feeding, 163
 method of feeding forages, 152
 milk quotas, 4, 178
 silage, 164
 straw feeding, 167

Mineral content, conserved forages, 44
Mixer-feeder wagon, 158
Moisture content (MC), effect on drying costs,
 15
 hay, 8
 of hay before loading, 96
 of silage, 16, 106, 168, 171
Molasses, silage additive, 19
Mowers, 64
 conditioners, 71
 height of cut, 76
 power requirement, 64
 working rate, 64
Mowing and conditioning, 64

National Institute of Agricultural Botany
 (NIAB), 28, 47

National Institute of Agricultural Engineering
 (NIAE), 31, 72
National Institute for Research in Dairying
 (NIRD), 163
Nitrogen, fertiliser, 56, 180
 fixation by legumes, 58
 non-protein nitrogen, 43, 145, 176
 T-sums, 57, 181
North Wyke Experimental Station, 58
Nutritive value of forages, 27

Outdoor clamp silos, 128
 walled silos, 125

pH, of silage, 17, 37
 in the rumen, 40, 44
Plastic sheets, for silo covering, 16, 120
 importance of weighting, 125
Propionic acid, hay additive, 13, 96
 silage additive, 23, 134
Protein value, crude protein (CP), 42
 non-protein nitrogen, 43, 145, 176
 of conserved forages, 12, 42, 173
 rumen-degradable protein, 42, 169, 173

Red clover, 34, 38, 55
Round bales, hay, 94, 150
 silage, 130, 160
 straw, 146
Rosemaund EHF, 182
Rowett Research Institute, 43, 170
Ryegrass, digestibility, 29, 48
 Italian, 35, 48, 55
Self-feeding, of silage, 152, 166
Self-loading forage wagon, 115
Sheep, feeding conserved forages, 174
 big-bale silage, 176
 indoor housing, 174
Silage, 16, 105
 acidity (pH), 17, 37
 buffer feeding, 181
 chemical additives, 19, 135
 crops, 55
 effluent, 18, 137
 digestibility (D-value), 32, 165, 171
 dry-matter (DM) content, 17, 106
 forage-box feeding, 156
 for beef cattle, 170
 for dairy cows, 163
 for sheep, 174
 high dry-matter, 106
 inoculants, 24
 intake, 38, 107, 152, 166, 176
 maize silage, 60, 134
 methods of feeding, 151
 minimum fermentation, 136
 proportion of conserved forage, 2, 178

protein content, 42, 173
sampling, 140
self-feeding, 152
storage, 139
unloaders, 154
Silage-making, 105
  big-bale silage, 130, 160, 176, 182
  big-bale wrapping, 132
  chemical additives, 19, 135
  chop-length, 107
  clamp and bunker, 128
  consolidation, 107, 123
  disposal of effluent, 18, 137
  Dorset-wedge method, 119
  filling the silo, 119
  harvesting, 108
  inoculants, 24
  need for additives, 25
  outdoor, 125
  soil contamination, 38, 119, 176
  surface wastage, 121, 129
  tower silage, 18, 138, 160
Silo, automatic unloaders, 154
  big-bale, 130
  clamp and bunker, 128
  Dorset-wedge, 119
  outdoor, 125
  plastic sheets for sealing, 16, 120
  polythene sleeve, 130
  tower, 18, 138, 160
  unloaders, 154
  walled, 125
Sodium bicarbonate, effect on rumen digestion, 41
Soil contamination, silage, 38, 119, 176
Storage feeding, 182
Straw, as animal feed, 141
  digestibility, 142
  processed, 143
    ammonia, 145
    feeding, 167, 176

oven method, 148
sodium hydroxide, 143
Sugar beet tops, for silage, 61
Sugar content, effect on silage fermentation, 17, 25
Sulphuric acid, silage additive, 19
Supplements, effect on forage intake, 40, 167, 172
  feeding to dairy cows, 41, 164
  feeding to beef cattle, 172
  feeding to sheep, 175
  methods of feeding, 157, 167
Swaths, drying, 63

Tedders, 78
Tower silos, 18, 138, 160
  chop length of crop, 139
  dry-matter content of crop, 138
  mechanised feeding, 160
  unloading, 139
Trailer, automatic tail-gate, 115
T-sums, 57, 181
Tunnels, for conditioning hay, 102

Urea, non-protein nitrogen, 43, 145, 176

Weather, dependence of conservation systems, 7, 19, 53, 81, 106
Welsh Plant Breeding Station, 29
West of Scotland College of Agriculture, 29, 165, 181
Wilting, 18, 38, 106, 168, 171
  by chemicals, 83
  dependence on weather, 19, 53
  effect on drying costs, 15
  effect on silage intake, 38, 168, 171
  field methods, 63
  minimum wilt for silage, 25, 107, 168
  silage effluent, 18, 137

Yield/digestibility patterns, 29, 47, 53